SHOPFLOOR MATTERS

Shopfloor Matters analyzes the changing institutional arrangements of shopfloor governance in twentieth-century American manufacturing and considers the impact of these institutional arrangements on shopfloor outcomes such as labor productivity and workplace health and safety.

Building on the work of labor historians, industrial relations scholars and institutional labor economists, the book offers not only a comprehensive analysis of the changing nature of shopfloor labor–management relations in the large manufacturing firms of this century, it also supplies empirical evidence of the effect of these institutional changes on labor productivity growth and injury rates. No other study has dealt with the broad sweep of shopfloor governance during the twentieth century, paid as careful attention to the processes by which shopfloor institutional arrangements have changed over these years, or offered hard evidence on the relationship between changing shopfloor institutions and changing outcomes.

The book begins with an analysis of the rise of company unions in the early twentieth century and the impact of company unions on labor productivity and workplace safety in the 1920s. Similar analyses, and related empirical findings, are offered for the rise of an empowered shopfloor voice for workers during the 1930s, the decline of workers' shopfloor empowerment during the late 1950s and early 1960s, and the rise of the recent shopfloor experiments with quality circles and team production in the 1970s and 1980s.

David Fairris is Associate Professor of Economics at the University of California, Riverside. He has published widely in professional journals on the subject of working conditions and shopfloor labor–management relations. This is his first book.

ROUTLEDGE STUDIES IN BUSINESS ORGANIZATIONS AND NETWORKS

SHOPFLOOR MATTERS

Labor–management relations in twentieth-century
American manufacturing

David Fairris

Routledge
Taylor & Francis Group

LONDON AND NEW YORK

First published 1997
by Routledge

2 Park Square, Milton Park, Abingdon, Oxon OX14 4RN
711 Third Avenue, New York, NY 10017, USA

Routledge is an imprint of the Taylor & Francis Group, an informa business

First issued in paperback 2016

Typeset in Garamond by Routledge

British Library Cataloguing in Publication Data
A catalogue record for this book is available from the British Library

Library of Congress Cataloguing in Publication Data
A catalogue record for this book has been requested

ISBN 978-0-415-12123-1 (hbk)
ISBN 978-1-138-98186-7 (pbk)

For Zak and Matt,
who matter even more

CONTENTS

ILLUSTRATIONS

FIGURES

TABLES

ACKNOWLEDGMENTS

This book began over a decade ago when, in a chapter of my doctoral dissertation, I suggested that the rise of company unions in early twentieth-century American manufacturing might be usefully viewed as an institutional change from "exit" to "voice" in the mechanisms by which workers expressed their shopfloor concerns to management. Since then—in between other research projects, a marriage, and two children—I have periodically returned to a study of the history of US shopfloor governance.

Sabbaticals at UC-Berkeley in the late 1980s, and at Harvard in the mid-1990s, gave me both the time and the inspiration to pursue these matters more seriously. During these sojourns, economists Lloyd Ulman, Michael Reich, Steve Marglin, and Richard Freeman helped to improve my arguments and to renew my interest in the subject. Long lunches with labor historian David Brody and industrial relations scholar George Strauss during the Berkeley sabbatical were also of great help to me.

Almost every part of this book has been presented at one time or another in various seminars and conferences across the country and beyond, and I thank the many participants who have helped to clarify my thinking on twentieth-century shopfloor matters. The economic historians at UC-Davis, Northwestern, and the University of Illinois; the labor economists and political economists at Harvard, UC-Berkeley, and Notre Dame; the radical economists at various conferences of the Union for Radical Political Economics; the labor historians at various conferences of the Southwest Labor Studies Association; the industrial relations scholars at various conferences of the Industrial Relations Research Association; and trade unionists at a conference of the Confederation of Venezuelan Workers, have all listened to me present my research for this book, and offered useful suggestions in return.

Many of the theoretical ideas that animate this study—the tension between shopfloor control and productive efficiency, for example, or the logic of institutional change—and many of the more sophisticated econometric findings produced during the course of researching this study—the time series regressions on the trajectory of postwar injury rates, for example—have been left for presentation elsewhere. I have also chosen to relegate to technical appendices the estimated regression coefficients for the empirical work presented in the book. Social scientists interested in such issues can read my other published papers, read between the lines of the book, and glance at the technical appendices.

It is my hope that what remains will be accessible to the broad audience of scholars interested in the history of the US workplace. I am particularly hopeful that labor historians will find value in the book. I have tried throughout the manuscript to document the enormous contributions labor historians have made to our understanding of this shopfloor history. As the project neared completion, and as I became increasingly uncertain about whether the book would indeed appeal to labor historians, I asked David Brody and Daniel Nelson to read what I had written. I am still uncertain about its appeal, but the book is a better one because of their comments and criticisms.

I gratefully acknowledge the permission to print herein portions of previously published articles. These include my papers: "From Exit to Voice in Shopfloor Governance: The Case of Company Unions," *Business History Review* 69 (Winter), 1995: 494–525, large portions of which appear in chapter 1; "Appearance and Reality in Postwar Shopfloor Relations," *Review of Radical Political Economics* 22 (4), 1990: 17–43, part of which appears in chapter 5; and "The Crisis in U.S. Shopfloor Relations," *International Contributions to Labour Studies* 1(1), 1991: 133–56, a large portion of which appears in chapter 7.

I received guidance and support, for which I am most grateful, from kind staff members at the Walter P. Reuther Library at Wayne State University; the Martin P. Catherwood Library at Cornell University; the Institute of Industrial Relations Library at UC-Berkeley; the Historical Documents Department at the Baker Library of the Harvard Business School; the Goodyear Tire & Rubber Company Archives in Akron, Ohio; and the National Archives. Many thanks also go out to the folks at the Saturn Corporation.

I have benefited from the research assistance of a slew of students over the years—most important were the efforts of Ranjeeta Basu, Mwangi Githinji, Wade Tang, Bassam Yousif, and Betsy Zahrt. I also had the especially good fortune of convincing Sandy Schauer to prepare the manuscript for publication. Sandy took a set of separate files on floppy disk that I confidently labeled "chapters" and turned them into a real book.

Finally, I made reference to my wife and two children above, but may have left the mistaken impression that they were mostly impediments to the timely completion of this project. Nothing could be further from the truth. Without Zak and Matt—and the seemingly endless soccer games and countless pets they introduce into my life—I would lack the emotional anchor that provides space for creative reasoning. I therefore dedicate the book to them. Although she did not spend the long hours in dusty archives, or suffer the countless frustrations involved in managing various data sets, or struggle with how to put down in print this or that idea, my wife seems like a co-producer of this book to me. And, in at least one respect, she truly is. One day I came home from the university with a number of high-falutin, scholarly sounding titles I had thought up for the book. Chris listened, looked a bit puzzled, and said, "Why not simply 'Shopfloor Matters'?" Why not indeed.

INTRODUCTION

In the winter of 1936–7, workers at two of General Motors' Fisher Body plants in Flint, Michigan quit work, sat down, and occupied the plants for over a month in what would become one of the most famous events in US labor history. The strike quickly spread to other GM facilities, eventually encompassing seventeen plants and idling 136,000 workers. Lost production was estimated to be roughly 280,000 automobiles, valued at $175 million. The strike ended with General Motors agreeing to recognize and engage in collective bargaining with the United Auto Workers of America (UAW), a relationship that would come to pit one of the most powerful corporations against one of the most powerful unions of the twentieth century.

This strike is remembered as the beginning of collective bargaining in the auto industry. It is often recalled that the strike began the year after Congress passed the National Labor Relations Act granting workers the right to organize unions and also after John L. Lewis broke with the skilled-trades dominated American Federation of Labor (AFL) to form the Committee for Industrial Organization in an attempt to organize the semi-skilled workers of the mass-production industries. The strike is commonly associated with the struggle for decent pay, health and pension benefits, and employment security, all of which would come to mark the achievements of collective bargaining in the auto industry and mass-production manufacturing more broadly in the decades that followed.

And yet in the minds of the workers who initiated the strike, shopfloor conditions were the primary cause and the single most important demand sustaining their efforts to organize a union. Discontent with the arbitrary and dictatorial behavior of foremen and, above all, the pace of production, animated their struggle. As Sidney Fine states in his careful study of the GM sit-down strike:

> It was the speed-up in the view of the principal participants that was the major cause for the GM sit-down strike. . . . It was the inexorable speed and the "coerced rhythms" of the assembly line, an insufficient number of relief men on the line, the production standards set for

individual machines, the foreman holding a stop watch over the worker or urging more speed, the pace set by the "lead man" or straw boss on a non-line operation, and incentive pay systems that encouraged the employee to increase his output.

(Fine 1969: 55)

William "Red" Mundale, one of the leaders of the sit-down strike, put matters most succinctly: "I ain't got no kick on wages," he said, "I just don't like to be drove" (quoted in Fine 1969: 57).

Roughly thirty years later, in the fall of 1964, a similar disruption occurred at GM, closing 89 of its 130 plants and involving roughly two-thirds of its workforce. B. J. Widick wrote in *The Nation*, "The nearly six-week shutdown . . . of General Motors must be described as the most prolonged and biggest 'wildcat' strike against American industry since the sit-downs of the turbulent thirties" (Widick 1964: 349). Although the national agreement between GM and the UAW—covering wages, fringes, and various employment securities—had been successfully hammered out, workers refused to return to work until pressing "local issues"—conditions at the plant level—were resolved.

This strike is remembered, like others during this period, as part of the " '60s rebellion." It is often noted that the average age of production workers in manufacturing fell significantly during the decade, and that African-Americans became more prominent in blue-collar production jobs during these years. The younger workers are thought to have possessed unrealistic expectations; they were less accepting of the authority of foremen and supervisors, and had not yet become acclimated to the rigors of mass-production manufacturing. Black-worker discontent, on the other hand, was part and parcel of the larger movement for racial equality and the elimination of discriminatory treatment at the hands of white superiors.

However, the " '60s rebellion" of blue-collar workers—and the 1964 "local issues" strike at GM in particular—was in large part the result of a growing expanse between the desires for a healthy, safe, and moderately paced work day and the reality of shopfloor existence. The strikes and other expressions of discontent did not involve younger workers exclusively; nor were these due to unrealistic expectations. Workers could point to the cold, hard fact that the quality of shopfloor conditions had been declining since the early part of the decade. Speedups were a common cry of workers over this period, and injury rates had begun to rise in manufacturing during the early 1960s after a sustained period of rapid decline following World War II.

Black workers bore a disproportionate share of the worsening shopfloor conditions. It was both this discriminatory treatment and the worsening conditions themselves that animated their struggle. The Dodge Revolutionary Union Movement (DRUM) and its affiliates in the auto plants around Detroit organized against discrimination with rhetoric and actions common to the black nationalist movement of the late 1960s, but, as Lichtenstein argues, "its industrial militancy targeted shopfloor issues virtually identical to those that had animated UAW radicals from one generation to the next" (Lichtenstein 1995: 433).

Georgakas and Surkin's account of the struggles of black "urban revolutionaries" in Detroit in the late 1960s and early 1970s concurs: "the movement led by black workers defined its goal in terms of real power—the power to control the economy, which meant trying to control the shop floor at the point of production" (Georgakas and Surkin 1975: 5). Perhaps the lyrics of a famous Detroit blues song by Joe L. Carter put matters most succinctly:

> Please, Mr. Foreman, slow down your assembly line.
> Please, Mr. Foreman, slow down your assembly line.
> No, I don't mind workin', but I do mind dyin'.
>   (quoted in Georgakas and Surkin 1975: ii)

Roughly thirty years later, in the spring of 1996, GM was once again faced with a virtual shutdown of its operations as the result of a strike by its workers. This strike began among 3,000 workers at two GM brake-parts plants in Dayton, Ohio, which supply 90 percent of the brake parts for the company's North American operations. Because of the "just-in-time" inventory system used by GM and many other modern manufactures—which economizes on costly inventories by producing parts just as they are needed—the seventeen-day strike resulted in the closure of twenty-six of GM's twenty-nine North American assembly plants, idling roughly 180,000 workers, and making it the largest strike at GM in over a quarter of a century.

Judging from the coverage of the strike in the business press, the issue was the outsourcing of jobs. GM had recently awarded an anti-lock brake contract to a nonunion plant of the Bosch Corporation in South Carolina, where average hourly labor compensation was well under one-half that at the Dayton plants (*Wall Street Journal* 1996b: 1). The strike was depicted as a classic dual between GM and the UAW, with the former trying to preserve its right to outsource jobs in an effort to once again become a "world class competitor" in the auto industry, and the latter desperately trying to respond to the declining rate of unionization among the auto-related workforce (*Wall Street Journal* 1996a: 1).

To the Dayton workers, however, health and safety concerns were at least as important as the outsourcing issue in their determination to strike GM.[1] Repetitive-stress injuries at the brake-parts plants had been on the rise over the preceding few years as management pressured workers to speed up production in an attempt to bolster the plants' competitiveness. Hardest hit were workers in the rivet and grind jobs of the drum-brake area. GM had committed to increasing the manpower for relief purposes in an attempt to address workers' accumulating grievances, but was slow to act on this and other safety-related promises. Following a month of particularly active grievance filing over health and safety concerns, Local 696 decided to strike to protest the company's failure to adequately address these shopfloor issues. Six hundred unresolved health and safety grievances existed at the time of the strike.

The decision by GM to outsource the anti-lock brake work was particularly depressing for the workers because it meant that, despite the heightened pace and rising injury rates, jobs were likely to continue to be lost in Dayton. Thinking about the uncomfortable tension between the two major grievances raised by the strike, Joe Buckley, shop chairman of Local 696, reflected, "I guess health and safety doesn't matter if we can't keep the jobs, but then again are the jobs worth keeping if they can't be made safe?"

A BRIEF ACCOUNT OF *SHOPFLOOR MATTERS*

As these accounts reveal, and as the following chapters will show in greater detail, shopfloor conditions matter to workers. The focus of this book is the shopfloor in twentieth-century American manufacturing. Non-shopfloor issues of importance to workers, and to a lesser extent the organizations that pursue these worker interests, are ignored almost entirely. In doing this, I do not mean to suggest that wages and fringe benefits are less important to workers than safety and a decent work pace, or that unions are secondary to shopfloor informal work groups for insuring distributive justice and the legitimacy of authority in the employment relation. I do, however, believe that these non-shopfloor worker goals and the organizations that pursue them have been unduly privileged in the history of the US working class in the twentieth century. In ignoring them, then, I mean both to emphasize the underemphasized and to right past wrongs.

I am concerned in this study with the institutional arrangements of shopfloor governance, by which I mean the formal rules and regulations as well as the informal customs and practices governing shopfloor labor—management relations and the determination of shopfloor conditions; with how such arrangements influence actual shopfloor conditions such as safety and the pace of production; and with the efficiency, legitimacy, and distributional consequences of shopfloor institutions. I am also interested in the process by which change occurs in these shopfloor institutional arrangements, and particularly with the role that worker, but also to some extent management, discontent plays in initiating this process of change. Although I try throughout to give a flavor of the diversity in shopfloor organization across firms and industries, my focus is on the "general" experience in manufacturing, and in the large mass-production firms in autos, rubber tires, meat packing, steel, and electrical equipment in particular.

In the 1930s, 1960s, and 1990s there was an extraordinary amount of

4

worker concern with shopfloor conditions, as suggested by our recounting of events in the history of General Motors above. These three periods represent transition years in the institutional arrangements of shopfloor governance. The 1930s and 1960s were periods of crisis in existing institutional arrangements, during which time there was significant deterioration in shopfloor conditions and militant expressions of worker shopfloor discontent. The 1990s is a period of institutional rebuilding following crisis, during which time labor and management are struggling to hammer out new institutional arrangements, the outcome of which will determine the distribution of shopfloor rewards—e.g., safety for workers versus productivity for firm owners—in the decades to come.

The 1930s was not the first period of transition in the institutional arrangements of shopfloor governance in twentieth-century American manufacturing. The first identifiable change in shopfloor governance came in the aftermath of World War I, when many progressive manufacturers adopted employee representation plans, or company unions. In chapter 1, I show how company unions arose as a mechanism for worker shopfloor "voice" in reaction to the inefficiencies of the existing "exit" approach, which held that if workers were unhappy with the level of safety, the pace of production, or how they were treated by foremen and supervisors, they should quit and seek work with a different employer.

The very high rate of worker quits in manufacturing during this period can be directly traced to workers' discontent with shopfloor conditions. But, this discontent bred more than just high labor turnover. Workers' sense of injustice in the share of shopfloor rewards going to employers, and their feelings of the illegitimacy of management's authority in production, fostered a lack of cooperation with management, high rates of absenteeism, low worker productivity, and occasionally more militant tactics such as organized slowdowns and even sabotage.

It was left to the budding personnel management movement of the 1910s and 1920s to point out to employers the costs associated with high labor turnover and low "worker morale." Employers began offering "welfare benefits"—e.g., pension plans and paid vacations—to employees, contingent on their length of service with the firm, in an effort to reduce turnover. Later, they instituted company unions as a solution to the morale problems. By substituting "voice" for "exit" as the mechanism by which workers influenced shopfloor conditions, turnover was reduced, labor–management communication regarding shopfloor production was enhanced, and those shopfloor conditions that could be profitably altered to workers' satisfaction were improved.

Chapter 1 offers empirical evidence to suggest that company unions served both to enhance labor productivity and to reduce injury rates in those manufacturing industries where they were most prominently employed. A case study of the Amoskeag Textile Mills in chapter 2 reveals that injury rates declined significantly following the emergence of the company union at this firm in the early 1920s.

The events of the 1930s, outlined in chapter 3, show that company unions contained the seeds of their own destruction. By schooling workers in the rudimentary elements of workplace democratic decision making, and by offering lessons in the basics of communication and negotiation to elected worker representatives, company unions had the unintended consequence of paving the way for workers' demands for a more empowered shopfloor voice. These demands emerged during the dramatic downturn in economic activity of the early 1930s. The depression created a context in which the quality of workers' shopfloor conditions deteriorated, thereby exposing the inadequacies of the 1920s company-union voice, and government policies became more sympathetic to the rights of workers to form independent unions.

Speedups were a constant source of worker discontent during the depression years, and the abusive behavior of foremen and supervisors was another common complaint. Workers' demands for a greater voice in production spurred the spread and further evolution of company unions across American industry during this period, but it was the industrial organizing drives of the late 1930s, under the banner of the Congress of Industrial Organizations (CIO), that ultimately led to workers' shopfloor empowerment. However, rather than the shopfloor structure of worker representation in the new industrial unions—which, in fact, was not very different from that of the evolved company unions of the 1930s—it was the organizing *activity* that produced the empowerment. In the process of organizing unions, workers built an informal shopfloor organization, composed of informal work groups and shop stewards, that produced enormous shopfloor influence.

Empirical evidence presented in chapter 3 suggests that workers' new shopfloor empowerment led to a significant reduction in the pace of work, as revealed by the strong negative association between labor productivity growth and the extent of unionization across manufacturing industries in the late 1930s. In chapter 4, I offer further evidence of the importance of workers' shopfloor discontent during this period, and of the relationship between this discontent and instances of worker militancy such as sit-down strikes, through a case study of labor–management disputes in meat packing. The case study evidence lends suggestive support to the claim that shopfloor empowerment emerged as much through the process of union organizing as by the attainment of union recognition or a collective-bargaining agreement with the employer.

Beginning during the early 1940s, and extending into the late 1950s, workers utilized a decentralized system of shopfloor governance—referred to in chapter 5 as "fractional bargaining"—to win sizeable improvements in shopfloor conditions. Shopfloor power existed in the new union structures at the level of workers' informal shopfloor organization. Newly empowered informal work groups operated in conjunction with shop stewards to exert great influence over day-to-day custom and practice in shopfloor decisions. This influence rested, in part, on the existence of decentralized management structures, in which foremen and lower-level supervisors were granted significant discretion in carrying out and even formulating company policies regarding production.

Injury rates in manufacturing declined precipitously in the decade following World War II. Workers in many manufacturing plants were also able to influence the pace of production through custom-and-practice struggles with shopfloor management over production standards and job descriptions. And yet production ran smoothly and productivity growth was rapid, presumably owing to workers' willingness to cooperate with management given their ability to ensure justice in the distribution of shopfloor rewards and the legitimacy of managerial authority in production. In chapter 5, I offer empirical evidence to suggest that injury rates decreased most rapidly during this period in those manufacturing industries where workers possessed the greatest shopfloor power. Rates of productivity growth, by contrast, were no different in these industries than in those where workers' shopfloor power—and therefore improvements in shopfloor safety—was much lower.

Fractional bargaining was a system of shopfloor governance with substantial benefits to society in the form of improved workplace safety and enhanced worker cooperation with management in production. However, the distribution of benefits from this system favored workers at the expense of employers. Beginning in the late 1950s, employers began systematically to reduce the scope of informal custom and practice in the determination of shopfloor conditions, centralize management decision making over such matters, and interpret labor's rights more narrowly as only those contained in collective-bargaining agreements. A strict "contract-and-grieve" approach to shopfloor governance—a system referred to in chapter 5 as "shopfloor contractualism"—emerged, granting management the upper hand in the determination of shopfloor conditions.

The immediate impact of these developments was a rise, beginning in the early 1960s, in the manufacturing injury rate and increased worker complaints of speedups in production. Empirical evidence presented in this chapter reveals that those manufacturing industries where workers had possessed the greatest shopfloor power in the immediate postwar period were significantly more likely to witness both rising injury rates throughout the 1960s and more rapid productivity growth in the early part of the decade. A case study of local contracts in the auto industry over the postwar period, presented in chapter 6, confirms the rise of the "contract-and-grieve" approach during the 1960s, as workers began demanding—but not necessarily winning—greater coverage of shopfloor conditions in local contract language.

Employers were unprepared for the extent of workers' discontent with the system of shopfloor contractualism, and especially with the reaction of workers—in the form of wildcat strikes, rising rates of absenteeism, and "work-to-rule" attitude towards production—to the worsening shopfloor conditions entailed in the new system. As the contract-and-grieve model revealed its extreme inadequacies in weak contractual protections and rising rates of unresolved grievances, labor's cooperation with management plummeted, along with workers' sense of the justice and legitimacy of the new arrangements. Labor productivity growth declined significantly during the late 1960s. Empirical work presented in chapter 5 reveals a direct relationship between these expressions of worker discontent and the productivity slowdown of this period. Moreover, those industries where workers had possessed the greatest shopfloor empowerment during the immediate postwar years were significantly more likely to witness a slowdown in productivity growth during this period.

In chapter 7 I show that employers began searching almost immediately for solutions to the systemic problems of shopfloor contractualism. Viewed in relation to past periods of transition, the latest one has been especially protracted. In the past decade, however, the pace of experimentation with new institutional arrangements of shopfloor governance has increased dramatically across many manufacturing industries. The vision that seems to have emerged from these experiments is an Americanized version of the Japanese "lean production" model, containing quality circles and team production, but few of the other accouterments—such as significant worker training and employment security—that characterize the Japanese approach to industrial relations.

Quality circles and team production represent a return to a more decentralized system of shopfloor governance, reminiscent of fractional bargaining, with a more immediate resolution of workers' shopfloor grievances and a level of worker input into shopfloor conditions that is greater than that which existed during the period of shopfloor contractualism. However, while the lean-production model eliminates some of those features of shopfloor contractualism that were most distressing to workers, it does not grant workers the ability to alter fundamentally the distribution of shopfloor rewards or contravene the authority of shopfloor management through increased empowerment in shopfloor decision making.

The impact of the lean-production model on labor productivity has been the topic of a fair amount of empirical research. A number of studies report mildly superior productivity in lean-production settings, but whether this stems from improved efficiency—due, for example, to greater labor–management cooperation—or mere speedups remains unclear. Other studies find little evidence of a significant effect on productivity. Existing case study evidence suggests that, if there exists a productivity boost, it may be due to a heightened pace of production.

The rise in manufacturing injury rates has continued relatively unabated since the 1960s. Empirical evidence presented in chapter 7 suggests that lean production contributed to the worsening health and safety record of manufacturing industries during the late 1980s and early 1990s. Injury rates increased more rapidly over this period in lean production plants, due largely to their commitment to total quality management techniques which may heighten the pace of production and put added stress on workers to engage in quality control. Cumulative trauma disorders increased less rapidly in lean production plants during this period, largely as a result of teams in production and ostensibly due to the freedom of team members to engage in job rotation. The results reveal, however, that the heightened injury rates in lean production plants far outweigh their relatively superior performance in reducing the growth of cumulative trauma disorders.

Chapter 8 contains a case study of the Saturn Corporation, including reports of personal interviews with workers at the Spring Hill, Tennessee, plant. Saturn's system of shopfloor governance goes far beyond the lean-production model in endowing workers with sizeable input into the shopfloor organization of production. On a host of criteria—product quality, labor–management cooperation, and worker satisfaction—Saturn appears to possess a system of shopfloor governance superior to that of the emerging lean-production model. However, the slow pace of production at Saturn has been a sore spot with GM for some time now, and because work pace translates into profit rates, GM has apparently decided to abandon the Saturn approach as a model for its other production facilities, and to put increasing pressure on Saturn's workers to boost their intensity of labor effort.

Several decades of rising injury rates and experience with a system of shopfloor governance that has proven unresponsive to workers' shopfloor concerns leads us, in chapter 9, to ask not whether lean production represents an improvement over the system of shopfloor contractualism, but whether it represents the best we can do. Shopfloor contractualism led to a growing sense among workers that the distribution of shopfloor rewards was unjust and that managerial authority in production was illegitimate. These, in turn, caused workers' cooperation with management in production to decline, and, along with it, the productive performance of firms. Restoring efficiency in production will require that workers' moral and political concerns be addressed, and thus that their shopfloor power be elevated.

The lean-production model that has recently emerged in the US does not grant workers a level of shopfloor power that will boost productive efficiency to the degree that is possible. Lean production is comparatively more efficient in Japan because in that setting it is combined with certain structural features—such as lifetime employment security and copious worker training—and norms of behavior—such as harmony and mutual obligations in labor–management relations—that generate just shopfloor outcomes and the legitimacy of authority in shopfloor governance. I conclude chapter 9 by suggesting that we therefore consider an alternative system of shopfloor

9

governance, one based on a works-council system of statutory shopfloor rights similar to that found in the German industrial relations system. Increased shopfloor empowerment of workers through a system of legislated works-council rights is a promising model for reform of America's system of shopfloor governance. It would restore justice and legitimacy in shopfloor arrangements, foster greater cooperation between shopfloor labor and management, and generate more efficient shopfloor outcomes.

THE CONTRIBUTIONS OF *SHOPFLOOR MATTERS*

This is the first study of its kind to focus exclusively on shopfloor conditions and shopfloor governance in twentieth-century American manufacturing. Perhaps its greatest contribution is the new light it sheds on the history of workplace safety and productivity growth in manufacturing, and the relationship between the changing trajectories of these two indicators of economic welfare and the institutional arrangements of shopfloor governance. However, this study also has implications for, by way of challenges to, widely held beliefs in the three disciplines that inspired the work: economics, labor history, and industrial relations.

Contributions for economists

Economists have a special fascination with how people and firms make choices among market alternatives, and with the efficiency of market outcomes. Until very recently, however, they have had little interest in institutions. Thus, there exists a naive tendency among economists to assume that economic analyses of the process by which people make market choices and the efficiency of their chosen outcomes can be applied equally well to institutional settings. For example, economists are inclined to view institutions as the intentional outcome of the rational decisions of self-interested individuals, and to see institutional arrangements as efficient by virtue of the invisible hand of competitive forces.

However, institutions pose special problems. At first glance, for example, the analysis of the transition from exit to voice in shopfloor governance seems to accord well with the economists' notion of rational choice and efficiency. The inefficiency of the exit approach to workers' expression of shopfloor discontent led firms to institute the more efficient voice approach, whereby company unions produced an improvement in the shopfloor rewards of both labor and management.

A careful analysis of the process of institutional change suggests, however, that the rise of company unions was contingent on their promotion by the War Labor Board during World War I. Moreover, both the War Labor Board and employers were responding to the growing demands of workers for a voice in shopfloor governance—demands which emerged as an unintended

10

consequence of employers' earlier attempts to reduce turnover through the provision of welfare benefits. Thus, although the move to a worker-voice mechanism for shopfloor governance was an efficient one, it could hardly be seen as either an inevitable or even intentional replacement for the less efficient exit mechanism. Unintended consequences and historical contingency may play a larger role in explanations of institutional outcomes versus market outcomes.

In addition, strategic factors—involving for example the distribution of jointly produced rewards—are arguably of greater significance in decisions concerning the formal and informal rules and regulations guiding the interaction among agents within institutions than such factors are in decisions concerning the purchase and sale of commodities. Cooperation between strategic actors also seems to play a more significant role in the efficiency of institutional outcomes versus market outcomes. And cooperation—unlike the decision to buy or sell something—depends, in a free and democratic society, on the distribution of rewards being seen as just and the delegation of authority being viewed as legitimate.

These factors are nicely illustrated in the shopfloor developments of the period since World War II. In eliminating fractional bargaining and initiating shopfloor contractualism, employers had hoped to undercut the power of workers' informal shopfloor organizations in order to improve their own distributive share of the shopfloor rewards. However, workers' feelings of injustice in the resulting shopfloor outcomes and illegitimacy in the authority granted management led to a significant decline in labor's cooperation in production, and thus a smaller pie for all to share.

The contemporary efforts of employers to transform the American workplace through introduction of the lean-production model from Japanese industrial relations offers similar lessons about the efficiency of institutional arrangements. While the American version of lean production might represent an improvement over the system of shopfloor contractualism, it is not the most efficient system available to us. There exist alternatives that would grant workers greater shopfloor power in production, thereby restoring justice and legitimacy in shopfloor outcomes and shopfloor governance, and thus efficiency in production. However, such a system might well have negative distributional consequences for employers, and so it is not on their agenda for possible adoption.

Contributions for labor historians

Students of labor history will also learn something of value from this study. There are two subject areas to which my analysis of the history of shopfloor governance is likely to contribute the most new insights: (1) company unions and (2) the rise of the industrial labor movement.

Shopfloor matters and company unions

The company unions of the early twentieth century are typically viewed either as part of the larger "welfare capitalism" movement of the period or as an attempt by employers to prevent workers from organizing independent unions. It is not entirely clear, however, why company unions came to be part of a package of employer-provided welfare benefits such as pension plans and profit-sharing arrangements. Similar unclarities exist in the claim that company unions served as a strategic move by employers to prevent the rise of independent unions. This claim makes more sense for the growth of company unions during the early 1930s than for the sizeable and ultimately stable and lasting company-union movement of the early 1920s, when workers' demands for independent unions had already been all but quashed. Moreover, there remains the question of how it was, exactly, that company unions served to quell workers' demands for independent unions.

A focus on workers' shopfloor discontents helps to eliminate some of these unclarities in the history of the company-union movement. The employer welfare measures of this period were meant, in part, to reduce intolerably high rates of voluntary quits by workers, many of which can be linked directly to workers' discontent with shopfloor conditions. However, having structurally reduced exits through welfare provisions, employers created a situation in which workers sought an alternative means—namely, voice—by which to express their discontent with shopfloor conditions. Company unions were the employers' answer to demands by workers for a voice in shopfloor governance. Thus, company unions became part of the larger package of welfare benefits because they acted as a voice substitute for the declining worker exits that welfare measures brought about.

Most workers no doubt would have preferred independent unions to company unions as a form of worker voice. Indeed, abrupt increases in union organizing activity occurred during the late 1910s and early 1930s, immediately preceding the two periods of most rapid growth in company unions. But how did company unions act to preempt workers' demands for independent unions during these years, and what explains the growth and staying power of the company-union movement of the 1920s? The answer to both of these questions can be found in a closer look at the impact of company unions on shopfloor conditions.

Company unions increased shopfloor safety and, in part through reduced injuries, boosted shopfloor labor productivity, thus benefiting both labor and management mutually. Company unions prevented the distributional losses to employers (and gains to labor) that independent unions would have produced—as implied by the conventional wisdom—but they were also a more efficient mechanism for shopfloor governance than the worker exit approach that preceded them, and so workers saw some benefits from their adoption as well. For workers, then, company unions may have been an inferior substitute for independent unions as a mechanism for the expression of worker interests, but they were nonetheless superior to relying on voluntary quits to accomplish this task.

Shopfloor matters and the rise of the industrial labor movement

A focus on shopfloor conditions, workers' shopfloor discontent, and institutional change in shopfloor governance also provides important insights into the events of the 1930s. The transition from company-union voice to an empowered voice in shopfloor governance for workers during the 1930s is one of the most fascinating transformations in twentieth-century American labor history, with several important implications for our understanding of the industrial labor movement and worker militancy during this period.

Scholars have not devoted sufficient energy to understanding the transition from company unions to independent unions. Conventional analyses of the rise of the industrial labor movement often give little credence to the role of company unions, thereby understating both their causal role in the rise of independent unions and the similarity of their formal shopfloor structure for dispute resolution with that of the new independent unions. Company unions provided the breeding ground for independent unions. They gave workers and their representatives skills in the collective representation of demands which were helpful in both building and sustaining independent unions. Moreover, many features of the new independent unions of the late 1930s—forms of shopfloor worker representation, formal grievance procedures, and even arbitration of disputes—can be found within the company unions of the mid-1930s.

Attempts to characterize the nature of the independent union movement of this period are also challenged by our shopfloor focus. One finds two extreme views in the literature: either the independent union movement is an example of "pure and simple" unionism, which demands a larger share of the rewards from production without fundamentally questioning the organizational structure and decision-making authority of management under American capitalism, or "revolutionary" unionism, which strives to overthrow the capitalist system of production and establish some alternative based on the political-economic principles of socialism or communism. The shopfloor focus of this study suggests that neither characterization fits very comfortably.

Most workers during this period possessed little desire to socialize ownership of the means of production. The movement was not "revolutionary." However, in their demand for an empowered voice in the determination of shopfloor conditions, workers were also seeking something far more ambitious than "pure and simple" unionism. The union movement, at least to the extent it genuinely represented workers' demands, stood for a fundamental change in the rights commonly attributed to those who own productive capital—namely, the right to control the conditions under which the capital equipment will be put to productive use. In contesting the control rights which are attached to ownership of the means of production, workers were making radical, if not revolutionary, demands.

It is, in part, the failure to acknowledge this shopfloor control aspect of the organizing efforts of workers over this period that has led to similar confusion over the militancy of workers' actions. Scholars have for some time now been trying to characterize the organizational goals of the often anarchic-seeming militancy of workers during this period. We now possess a long list of organizational aims that this worker militancy was not striving to achieve. The rank-and-file militancy was not directed at the attainment of socialism; nor at true "workers' control" like the movements of craft workers in the late nineteenth and early twentieth centuries; nor at the adoption of a strong shop steward movement similar to the one that developed in Britain during World War I; nor, for that matter, at building stable union structures, since militant workers were unlikely to have accomplished this without the efforts of committed CIO leaders.

Our shopfloor focus suggests that workers did not know what form of organizational voice they wanted or needed, but they understood that in the process of organizing independent unions they had arrived at an in-plant form of local empowered shopfloor voice that produced substantive improvements in shopfloor conditions and which deserved to be respected by those union leaders whose stated sympathies were with the workers. This informal shopfloor organization was not revolutionary, nor did it lead to demands for "worker control" like that of nineteenth-century skilled workers, but it was not a simple desire for "the workplace rule of law" which would come to characterize the labor leadership's approach to worker shopfloor voice during the post World War II period. Workers' initial reticence about CIO organizing was premised on skepticism about the ability of such organizations to influence shopfloor conditions. Workers' later commitment to CIO union structures waned only when the stability of union organization appeared to demand that workers forgo winning shopfloor improvements through informal, and sometimes militant, shopfloor actions. And, finally, a closer examination of workers' informal shopfloor organizations during this period may lead to the conclusion that they bore more of a resemblance to the British shop stewards' movement than is generally acknowledged.

Contributions for industrial relations scholars

Students of postwar industrial relations will also learn something from, or at least be challenged by, this study of shopfloor governance. There are two topic areas to which this analysis contributes important new insights: (1) the postwar industrial relations system; and (2) the contemporary period of workplace transformation.

14

The postwar system of industrial relations

The conventional interpretation of industrial relations during the quarter century following World War II is that by the late 1940s, labor and management had come to an agreement about both the issues that would become subject to joint regulation under the new union–management relationship (from among the list: wages, hours, and other conditions of employment) and the institutional arrangements to be used in jointly regulating them (primarily collective bargaining and the grievance procedure). Having done so, the stage was thus set for a lengthy period of industrial stability, mutual prosperity, and rapid economic growth. Where, and in what way exactly, shopfloor conditions fit into this view of the postwar industrial relations system remains extremely vague.

Some accounts seem to assume that shopfloor issues such as the pace of production became as amenable to joint contractual regulation during this period as did wages and hours. Other accounts appear to view employers as having demanded "managerial prerogative" over shopfloor matters such as work pace, both as a condition of union recognition and in exchange for granting workers generous wage and fringe benefits in collective-bargaining agreements. However, a careful analysis of shopfloor governance during the postwar period suggests that neither account squares particularly well with reality.

Workers did not turn to contractual protections and formal grievance handling during the immediate postwar years to achieve their shopfloor goals. Nor did they abdicate control of the shopfloor to management. What conventional accounts ignore is the custom and practice of shopfloor governance during this period. Focussing on informal shopfloor institutional arrangements, one sees a period of roughly fifteen years following the war during which time workers possessed sizeable influence over the contested terrain of shopfloor production. This was followed by a period, beginning in the late 1950s and early 1960s, when workers' informal activities were severely curtailed and workers' shopfloor influence was limited largely to the contractual guarantees they were able to establish in collective-bargaining agreements. It is during the latter period that the conventional accounts of a "contractualist" regime begin to ring true.

This new view of postwar shopfloor governance does more than simply render more accurate our characterization of shopfloor industrial relations during the postwar period to the early 1970s. It also sheds light on the events of the 1960s—such as the rise in manufacturing injury rates, the worker militancy, and the productivity slowdown of the latter half of the decade—in a way that more conventional analyses do not.

15

The contemporary period of workplace transformation

This study also poses a challenge to conventional interpretations of the contemporary period of US workplace transformation. Industrial relations scholars typically point to technological change, rising international competition, and innovations in the theory of human resource management as causal factors in the current efforts of employers to transform their workplaces. In most analyses, shopfloor matters do not appear on the list of factors contributing to the workplace transformation movement. This is rather curious, however, in light of the fact that institutional changes in shopfloor governance—quality circles and teams, for example—have played such a large role in the recent reform efforts.

Our analysis suggests that when domestic employers began instituting quality circles, work teams, and other aspects of the lean-production model during the 1970s and 1980s, they were searching for solutions to an existing crisis in shopfloor governance as much as they were attempting to imitate the industrial relations systems of international competitors or engaging in organizational adaptation to new technologies. The system of shopfloor contractualism led to a deterioration in the shopfloor conditions of workers, and thereby growing worker discontent and rising labor militancy, the results of which were declining worker cooperation with management and the productivity slowdown of the late 1960s.

The goal of the contemporary workplace reform movement, as conventionally understood, is to alter postwar institutional arrangements so as to foster greater cooperation between labor and management and enhance labor's productivity in production. However, failure to ground their analysis in the crisis of shopfloor contractualism has left many conventional analysts without a clear vision of the types of changes that are required to truly boost labor–management cooperation and productive performance. An analysis focussed on the shopfloor and the history of shopfloor governance suggests that this will require restoration of workers' lost sense of justice in the distribution of shopfloor rewards and legitimacy in the authority granted to management. This, in turn, will require the substantive empowerment of workers in shopfloor decision making.

Which, in a rather sad sort of way, is also precisely where our story begins.

1

FROM EXIT TO VOICE IN SHOPFLOOR GOVERNANCE

Factory workers in late-nineteenth-century America possessed numerous shopfloor discontents: cleanliness was at a minimum, lighting was horribly inadequate, ventilation was poor, and the temperature was dictated by either the nature of production or the season. Each industry had its special problems. Work in the iron and steel mills was hot, dirty, and physically exhausting. Meat packing work was cold and wet, with floors so filled with slime and grease that moving about the factory floor was treacherous. The foundries of metal-working plants were dark and dank, with bitterly cold temperatures in the winter, except, of course, at casting time when the heat was almost insufferable.

Technological and organizational developments which took place around the turn of the century led to improvements in some shopfloor conditions. Mechanization was often attributed with reducing the physical strength required for a day's work, as happened for example in steel production. And the new, larger plants of the early twentieth century were generally acknowledged to be cleaner, better lighted, and better ventilated than the older factories. Conditions of production also improved as a result of conscious efforts at progressive reform. By the turn of the century almost every northern industrial state had passed legislation requiring that factories be clean and well ventilated, and that safety guards be used on dangerous equipment (Nelson 1975).

However, the very same technological and organizational developments that brought forth improvements in some shopfloor conditions led to deterioration in others. Among the worst of the worsening conditions were the capricious and dictatorial behavior of foremen, the heightened pace of production, and declining health and safety.

The economies of scale associated with the new technology and organization of production led to the construction of larger plants. Apart from the textile mills, factories with over 500 wage earners were virtually nonexistent prior to the 1870s. By 1900, there were roughly 1,500 factories of such size, almost a third of which contained over 1,000 wage earners (Nelson 1975: 4). Growth approached epic proportions during the next two decades, so that by the early

1920s plants employing at least 20,000 wage workers were rather common-place in such industries as steel, autos, electrical equipment, and rubber.[1]

The newer plants may have been better lighted and ventilated than the older ones, but the increased size of operation also placed the bulk of workers' supervisory contacts in the hands of foremen instead of factory owners. And the former wielded shopfloor power in ways that did not always translate directly into either profits or worker welfare. Especially offensive to workers was the use of arbitrary criteria for assigning workers to tasks and in recommending workers for promotions, and the dictatorial zeal with which foremen drove workers to produce.

The "drive system" of production which foremen enforced, often through tactics of intimidation and brutality, produced yet another source of worker discontent during this period—a heightened work pace (Slichter 1919). Mechanization was a contributing factor as well in that, although it may have made work less physically demanding, it also allowed for an increase in the overall speed of production (Chandler 1977). The increased pace of production, and its impact on the worker, was noted in a 1892 report from the Maine Bureau of Industrial and Labor Statistics:

> Work is not done in the old, slow way, and in nearly all industries, by the present methods from two to four times the quantity of product is turned out in the ten hours. How much faster is the operative compelled to work, and how much greater is the strain, to accomplish this amount of work, in comparison with the old twelve-hour method?
>
> (quoted in Atack and Bateman 1991: 5)

Cross-country evidence on the pace of production in early-twentieth-century textile manufacturing corroborates the claim of a breakneck speed in US plants. Clark (1987) found that accounting for differences in training, technology, managerial practice, and the like, one New England cotton textile operative in 1910 did 6.0 times the work of the equivalent Greek, Japanese, Indian, or Chinese operative; 2.3 times the work of a German operative; and 1.5 times the work of a British operative. It was generally conceded by foreign visitors that the level of work intensity was significantly greater in most American factories compared to similar factories in their own countries.

Health and safety issues also became more prominent during this period. Heavy machinery produced near-deafening noise levels and ceaseless vibra-tions. Fast-moving belts and gears required constant attention by workers if they were to avoid accidents. The increased use of chemicals in the work-place raised important health concerns. And the new metal and mineral grinding and polishing equipment produced a fine dust that scarred lung tissue, providing the perfect breeding ground for a variety of respiratory infections. (The tuberculosis bacillus alone was responsible for perhaps half the deaths of industrial workers in the late nineteenth and the early twentieth century (Atack and Bateman 1991: 5).) The noise and vibrations of the equipment in combination with the hectic pace of production put great strain on both mental and physical endurance, producing a form of fatigue that also threatened health and safety in many plants.

National statistics on industrial injuries and fatalities were not systematically gathered until the mid-1920s. However, a study of workplace safety in the iron and steel industry during the early twentieth century conjectured that the accident rate in all of manufacturing was probably higher between 1903 and 1907 than at any other time in US history (US Department of Labor 1918: 13).[2]

Private studies of workplace health and safety also give some clue as to the extent of the problem during these years. Eastman's study of industrial safety in the Pittsburgh area during twelve months of production in the years 1906–7 found accidents in such industries as railroading, steel, and mining resulting in 526 fatalities and 509 injuries requiring at least one night's stay in the hospital (1969: ch. 1). The iron and steel industry had the worst record of fatalities. Interestingly, though, there was significant variation in safety performance within the industry. The Carnegie Steel Co. had a fatality rate of 32 per 10,000 workers, while Jones & Laughlin had a rate of 56 fatalities per 10,000 workers (Eastman 1969: ch. 4). A US Bureau of Labor study of the iron and steel industry in 1909–10 concluded that nearly one-quarter of the full-time workforce suffered an injury in the 155 plants studied (Nelson 1975: 30).

Eastman's study attempted to go beyond the mere documentation of accidents in order to arrive at the factors responsible for them. Roughly 30 percent of the accidents for which responsibility could be reasonably established were attributable to employers or their representatives in positions of authority; and many of these, the study argued, were preventable. For the 28 percent of accidents caused by worker carelessness or inattention, the study pointed to "the long hours of labor" and "the high speed and unremitting tension" as the ultimate determining factors in many of them. The study concluded:

> we have much deliberate disregard for safety in the construction of plant and equipment, and in the organization of work; we have found a long list of defects . . . most of which careful inspection would have revealed and immediate repair have rendered harmless; we have found those directly representing the employer in positions of authority often neglectful of safety.
>
> (Eastman 1969: 103)

THE PERIOD OF WORKER EXITS

Workers looking for a way to express their discontent with the state of shopfloor conditions during this period found very few outlets. In 1909, F. N. Hoffstot, then President of the Pressed Steel Car Company, remarked, in reference to workers' ability to influence their conditions of employment, "If a man is dissatisfied, it is his privilege to quit" (quoted in Brody 1960: 78). A more succinct description of the primary means by which workers—and less-skilled workers especially—expressed shopfloor discontent during these years cannot be found.

No systematic surveys of labor mobility exist prior to 1910. Thus, our knowledge of worker turnover before this time is industry or plant specific and largely anecdotal. In 1906, for example, Commons found a machine shop in Pittsburgh which had to hire 21,000 men and women in one year just to keep a workforce of 10,000 (Commons *et al.* 1935: 331). In 1907, ninety-one southern textile mills hired 57,000 new workers, but at no time had more than 30,000 employees on their payrolls (Nelson 1975: 86). Records from a steel mill reveal that for the years 1905–7, the number of newly hired workers was roughly 90 percent of the labor force (Nelson 1975: 86).

More systematic surveys exist for the 1910s. A prewar survey of firms by the US Bureau of Labor Statistics (BLS), soliciting information for the entire period 1910–15, revealed that during a twelve-month period in the years 1913–14, eighty-four establishments with roughly 245,000 full-time workers had an average separation rate (i.e., the total number of quits, discharges, and layoffs relative to the workforce of all establishments) of 100 percent (Brissenden and Frankel 1920: 41).[3] Voluntary quits were by far the largest component of the three types of separations, accounting for 76 percent of separations in 1913 and averaging roughly 70 percent for the entire period 1910–15. The study also revealed that separation rates were much higher (over double) for less-skilled workers compared to skilled workers, and for workers with less tenure with the firm (over 80 percent of the separations in 1913–14 occurred among the group of workers with less than one year's service with the firm).[4]

Slichter's (1919) study of labor turnover in 105 factories over the years 1912–15 revealed annual separation rates ranging from 348 percent of the total workforce in one factory to only 8 percent in another; the average, however, was nearly 100 percent (Slichter 1919: 22).[5] Almost 40 percent of the plants surveyed had annual turnover rates exceeding 100 percent. Slichter's survey results also supported the view that voluntary quits dominated separations, that turnover decreased significantly as the skill of the worker increased, and that separation rates were greatest among workers with less tenure with the firm (Slichter 1919: 44, 57–74, 85–9).

One of the distinctive features of Slichter's study was that he attempted to uncover the causes of worker turnover. Slichter discussed a number of factors leading to the high rates of voluntary quits in particular, including obvious ones like inadequate pay and lack of promotion possibilities. But he dwelled at some length on disagreeable working conditions as an important factor, including work pace, features such as dust, heat, monotony and nervous strain,

and conflicts between workers and foremen. Slichter offered numerous anecdotal examples, and occasional hard documentation, suggesting that quit rates
were especially high in foundries (because the work is hot, heavy, and dirty), in
sandblasting (because of the dust), in cleaning and varnishing departments
(because of the strong chemicals and the odor), and in many jobs where the
physical strain, due to the work pace and general exhaustion, was great.

Prior to the technological and organizational developments of the late
nineteenth and early twentieth centuries, employers' reliance on quits as the
mechanism by which less-skilled workers expressed shopfloor discontent
may have made sense. However, this approach became less sensible for
employers with the emergence of the new technology and organization of
production. The use of less-skilled workers, as a proportion of the labor
force, increased under the new techniques, portending an overall increase in
factory turnover rates. The growth was most dramatic in the large mass-
production industries. Statistics on semi-skilled operatives from the Ford
Motor Co., for example, reveal that as a percentage of total production-
worker employment, semi-skilled operatives increased from 29.5 percent in
1910 to 62.0 percent in 1917 (Gordon *et al.* 1982: 133).

An increase in plant-wide labor turnover, due to the increased use of
turnover-prone workers, was compounded by an increase in the average cost
of worker turnover during this period. The increased use of specialized
equipment, and the variation across firms in the form and extent of the division of labor and specialization, meant that the job tasks of the semi-skilled
workforce became increasingly specific to a single establishment or small set
of establishments. Firms bore a larger share of the training costs associated
with these firm-specific skills, and therefore lost a significant investment in
the skills of a worker if that worker were to quit. As the less-skilled contingent grew in proportion to the rest of the workforce, and as the jobs they did
required more on-the-job firm-financed training, exits became costlier to
employers (Doeringer and Piore 1971).[6]

Many employers were initially unaware of the costs associated with the
high labor turnover. The growth in plant size and the widespread absence of
systematic policies for human resource management meant that many
upper-level managers were often ill informed about such issues. Indeed, as
we shall see, among the significant contributions of the personnel-
management movement during the 1910s and 1920s was to impress upon
management, sometimes with exaggerated claims, the productivity and
production costs consequences of worker quits (Jacoby 1985). The larger,
more progressive employers would increasingly come to focus their attention
on this issue during the 1910s, attempting through various means to bind
workers more closely to the firm.

Workers actively seeking to develop an alternative to the exit mechanism
for expressions of shopfloor discontent faced a number of disincentives prior
to the 1910s. Less-skilled workers' labor-market prospects were greatly

enhanced by the technological and organizational changes of the period, and turnover was an important means by which to attain better pay if not better conditions. Especially during periods of tight labor markets, "workers improved their incomes as much by moving from job to job as they did by striking," notes Montgomery (1979: 96) of the period preceding the war. Moreover, for those immigrant workers who hoped one day to return to "the old country," amassing a significant sum of money was the all-important goal of their stay. Efforts to establish an alternative mechanism such as voice for expressions of shopfloor discontent could at best lead to improvements in shopfloor conditions that many considered to be only temporary.

Another major impediment to the formation of a voice mechanism for expressions of shopfloor discontent was the classic free-rider problem. Many of the shopfloor conditions of concern were essentially local public goods for the workers in a specific department or plant. Of course, certain shopfloor conditions—such as temperature, lighting, and general sanitation—had always been collectively "consumed" by workers. However, with the spread of mechanization and organizational changes such as the assembly line, production became more integrated in the factories of the early twentieth century, the result being that a larger number of shopfloor conditions became collective goods for a larger number of workers.

When shopfloor conditions are local public goods, no individual worker can significantly alter the speed of the line, the repetitive nature of the job, the illegitimate authority of the foreman, or the unhealthful shopfloor environment without those conditions changing for fellow workers as well. Moreover, once altered for one worker, other workers can benefit without significant additional costs. Whether worker voice was to be merely a mechanism for collectivizing workers' preference over local public goods, or a more ambitious attempt to alter significantly the distribution of shopfloor rewards and the illegitimacy of managerial authority, it would have to be *collective* in form.

Many of the impediments to the development of a collective worker voice mechanism facing employers and employees during this period would be significantly reduced by the events of the 1910s.

THE TRANSITION FROM EXIT TO VOICE

Slichter's (1919) study of turnover in the US labor market in the 1910s expressed the view that worker quits were costly to employers, that poor shopfloor conditions were in part to blame for these expressions of worker discontent, and that a plant-level mechanism by which workers could voice their shopfloor concerns might, if taken seriously by employers, lead to both better conditions and enhanced productivity. In contemporary parlance, Slichter was advocating a transformation from "exit" to "voice" in the governance of shopfloor labor–management relations (Hirschman 1970; Freeman and Medoff 1984). Instead of quitting in response to poor shopfloor conditions, workers might express their concerns about those conditions directly to employers, and in doing so provoke a mutually advantageous adjustment in shopfloor outcomes that could not be accomplished through worker quits.

In fact, the transition from exit to voice in shopfloor governance had already begun by the time Slichter's analysis of the inefficiencies of high labor turnover had been published. The transition began in the early 1910s with employers' increased awareness of the costs of labor turnover. Alexander's pioneering study (1916) of turnover in the metal-working industries during 1912 became emblematic of the research findings of the budding personnel-management movement of the period; its conclusion was that worker quits were both significant and costly to the firm. The search for worker replacements, combined with lost production, inferior products, increased wear and tear on equipment exacted by workers unfamiliar with their work, and resources devoted to worker training, amounted to a sizeable expenditure for the firm. Alexander (1916: 128–44) calculated that overall replacement costs for workers at General Electric were roughly $8.50 for a laborer, $48.00 for a skilled mechanic, and $73.50 for a semi-skilled operative.

Quits, however, were only one manifestation of workers' shopfloor discontent, and not even the most significant in the view of some employers. A personnel-management textbook from the period suggests that

> management is interested in labor turnover not so much from the point of view of the cost of replacing the men who leave, as it is interested in labor turnover from the point of view of lessened interest and effectiveness throughout the organization.
>
> (Jacoby 1985: 126, footnote 63)

Workers saw the "drive system" as an illegitimate form of managerial authority. They looked upon the pace and safety hazards of the new regime of production as a gross injustice in the distribution of shopfloor rewards; labor's interest in safe work at a reasonable pace was being sacrificed for management's interest in high labor productivity. Consequently, workers were less than diligent in their duties and less than devoted to their employers. Absenteeism was high, "soldiering" was significant, and sabotage was not uncommon as a result of this "worker morale" problem in production (Lazonick 1983).

A variety of approaches were developed by employers during this period to reduce the high rate of voluntary worker quits and to address the problem of low worker morale. Perhaps the most famous was the approach taken at the Ford Motor Company, where the turnover rate in 1913 was 370 percent, and absenteeism was averaging 10.5 percent per day. Ford's near doubling of the wage to $5.00 a day in 1913 for workers with at least six months service with the firm produced a dramatic decline in the rates of turnover and absenteeism, to roughly 40 percent and 0.4 percent, respectively (Lazonick

1983: 122). Despite its fame, this approach was not widespread among employers of the period.

Another solution lay in the development of internal labor markets (Doeringer and Piore 1971; Elbaum 1984), which offered employers a way of economizing on the costs of worker training and reducing labor turnover, at the same time that it provided workers with greater employment security and promotion possibilities within the firm. An internal labor market is a mechanism for recruiting workers for job openings from within the firm as opposed to going to the external labor market to fill positions. In an internal labor market, jobs are organized into distinct job ladders, with each rung of a given ladder representing a job that requires incrementally greater worker skills than the job rung immediately below it.

With an internal labor market in place, workers have an incentive to remain with a firm and ascend its job ladders, progressively attaining greater skills, job security, and pay as they proceed. Quits should therefore decline and job skills should be imparted in a more efficient manner. Internal labor markets appear to have spread to the ranks of the semi-skilled from the ranks of skilled workers during this period, but how widespread they were and how lasting were employers' commitments to them remain unclear (Jacoby 1985).

Another solution to the problems of worker turnover and morale—one adopted by many firms during this period—was to bind workers financially to the firm and increase workers' sense of loyalty to the enterprise through various forms of employer paternalism. These were the central commitments of the "welfare-capitalism" movement of the 1910s (Brandes 1970). In a series of articles in 1917, General Electric's public relations officer, George M. Ripley, described the welfare measures GE had recently undertaken to create "steady workers." The great bulk of these measures constituted an array of fringe benefits tied to length of service. Bonuses were available to employees with five years of service to the company, a week of paid vacation to those with ten years' service, and a pension plan for those workers with twenty years' service. In 1917 almost half the workers at the Schenectady plant and 35 percent of those at the Lynn plant qualified for the five-year bonuses.

Goodyear Tire & Rubber Company was another leader in the welfare-capitalism movement. In 1914, paid vacations were extended to factory workers, with one week's paid vacation going to workers with at least five years of service and two weeks to those with at least ten years of service (Allen 1949: 174). A pension plan was introduced in 1915, granting workers with at least fifteen years of service a pension beginning at age 70, the amount to be based on a percentage of total earnings during the years of continuous service to the company.[7] In mid-February 1918, a highly touted stock option plan was introduced. In addition to giving workers the chance to purchase company stock and receive dividends, the plan granted worker-shareholders payment of $3.00 per share per year for the first five years if they remained employed with the company and showed "a proper interest in its welfare."[8] By early March of the same year, more than $1 million in stock had been sold to factory workers alone.[9]

The increase in voluntary turnover during the early war years forced many employers who had not yet taken note of the consequences of workers' shopfloor discontent to do so. Schemes to reduce turnover and foster worker loyalty similar to those at GE and Goodyear flourished during this period. New welfare programs emerged in many firms that had not experimented with such benefits prior to the war, while firms that had directed their earlier welfare measures towards the reduction of turnover among skilled workers began to extend them to the less skilled as well (Brody 1960).

In some cases, these welfare programs were combined with efforts to improve personnel management through the creation of personnel departments which would oversee the process of hiring and firing, introduce systems of internal promotion based on internal labor-market principles, and even regulate the driving behavior of foremen. The personnel-management movement, which emerged just before the war, was initiated by groups of engineers interested in human-relations issues, and welfare workers and vocational guidance experts interested in issues of plant efficiency. Jacoby notes that between 1915 and 1920 the proportion of large firms (over 250 employees) with personnel departments increased from 5 percent to 25 percent (Jacoby 1985: 137). These early personnel departments were faced with two impediments which tempered their efforts to address workers' shopfloor concerns directly: lacking a mechanism for communicating with workers, they were ill informed of the precise nature of workers' shopfloor discontents, and they confronted the powerful resistance of foremen in attempts to ameliorate those conditions for which foremen were responsible.

The early efforts by management to reduce turnover, on the other hand, appear to have been somewhat successful. Labor turnover was very high during the war years to be sure, but this was due in large part to the extraordinarily tight labor markets created by a combination of the new immigration policy, war-time conscription, and increased orders in those indusries intimately connected with the war effort. Brissenden and Frankel's findings on aggregate labor separations in 1913–14 and 1917–18 reveal rates (per 10,000 labor hours) of 3.3 and 6.7, respectively (Brissenden and Frankel 1920: 44). However, the unemployment rate was 7.9 percent in 1914, but only 1.4 percent in 1918 (US Bureau of the Census 1975: 135).

The increase in separation rates varied widely across manufacturing industries at this time: metal products manufacturing witnessed a threefold increase in separations between 1913–14 and 1917–18, but separations in the auto and auto parts industry increased by less than 50 percent, and the rate in slaughtering and meat packing actually fell (Brissenden and Frankel 1920: 44). Were these across-industry differences in the trajectory of separation rates due, at least in part, to differences in the extent of welfare provisions?

One piece of evidence suggesting this may have been the case comes from an analysis of survey evidence on welfare measures and their impact across establishments gathered by the BLS in 1916–17. The survey asked establishments about the extent to which they offered various kinds of welfare benefits to employees, and, later in the survey, whether employers thought their welfare efforts had produced a discernible impact on turnover. These survey results made possible the creation of a "welfare index" capturing the extent of welfare provisions across ten manufacturing industries and an industry-level measure of the degree of success establishments reported with "reduced turnover" owing to their welfare benefits.[10] The simple correlation between the industry "welfare index" and the measure of "reduced turnover" was positive and significant at the .10 level, offering at least suggestive evidence in support of the claim that welfare benefits reduced voluntary separations to levels lower than they otherwise would have been during the late 1910s.[11]

To the extent that the efforts at turnover reduction were indeed successful, they had unintended consequences. Having succeeded in raising the costs of exits for workers, firms laid the foundation for voice as a substitute mechanism for expressing shopfloor discontent (Hirschman 1970). Reduced turnover served a dual role. Firstly, with less recourse to the exit option for job matching, workers required an alternative mechanism for conveying preferences over collectively experienced (and, during this period, constantly changing) shopfloor conditions.[12] Secondly, a more stable workforce was in a better position to utilize "selective incentives"—such as social ostracism—for overcoming the free-rider problem which typically plagues movements to achieve collective goals (Olson 1971). The result was increased demand for a collective voice in the determination of shopfloor conditions.

Uprisings of less-skilled workers had emerged in opposition to the stretch-out (an increase in the number of machine loads a worker is expected to tend) in the textile industry and to the arbitrary behavior of foremen in the clothing industry in the early 1910s (Fraser 1991). This situation repeated itself across many of the major manufacturing firms in the latter part of the decade. The number of work stoppages in 1916 was unprecedented, and would not again be witnessed until 1937 and the successful CIO organizing drives of that year. The average number of work stoppages annually over the years 1916–20 would not be witnessed again until the World War II period (Lazonick 1990: 244). Data on national strike activity reveal the growing number of strikers in relation to strikes as the decade progressed, reflecting the increased support and involvement of the less-skilled workers of the day. In 1919 one out of every five employed workers was out on strike at some point during the year (Brody 1965: 129).

Shopfloor conditions such as pace, safety, and the arbitrary behavior of foremen were prominent concerns in the strikes of this period. As always, pay and hours of work were also central concerns, but even these were often intimately related to shopfloor conditions, as when workers complained of bonus-payment schemes that intensified the pace of work or demanded hours reductions because of fatigue and the enhanced risk of injury it engendered. According to Montgomery, the years 1916–20 represent one of three time periods since the 1880s that US workers have engaged in extensive struggles with employers over issues of "control" in production (Montgomery 1979: 98).[13]

The list of firms witnessing major strikes over this period reads like the list of firms that had been at the forefront of the welfare-capitalism movement. Strikes occurred at the McCormick Works of International Harvester in 1916, at GE's Lynn plant in 1918, and in steel in 1919. However, it is possible to marshall evidence of a less anecdotal nature suggesting the existence of a causal relationship between increased structural impediments to worker exits due to welfare benefits and increased demands for worker voice during this period.

An industry-level measure of the growth in strike activity—"strike growth"—over the latter half of the 1910s was created from published sources, and its correlation with the industry-level variable "reduced turnover" was explored statistically for six of the manufacturing industry groups.[14] The simple correlation between the two measures was positive and significant at the .10 level of significance, offering at least suggestive evidence for the claim that those industries with the greatest success at reducing turnover through welfare benefits also witnessed the largest growth in strike activity during the late 1910s.

If the demands for worker voice emerging during this period were in part an unintended consequence of the disincentive to exit provided by welfare benefits, they were also in part the result of historical contingency. The contingent force was World War I. Several features of war-time production gave greater impetus to workers' demands for a voice in industry while simultaneously shaping the outcome of these demands. The tight labor markets brought on by war-time production meant that workers could act on their desires for a voice in determining shopfloor conditions with greater confidence of landing alternative employment should they be dismissed for their activities. (Dismissals could prove terribly costly for workers, especially since negative referrals and black-listing were widely practiced at the time.)

Immigration restrictions during (and after) the war reduced the ease of mobility back and forth from the "old country." Therefore, the poor shopfloor conditions that before had been considered merely temporary features of a life's work came to be viewed as permanent. Meanwhile, war-time production brought a general deterioration in many of these same shopfloor conditions. Accident rates rose rather dramatically, the speedup was deployed with greater intensity, hours of work increased across many industries, and ramshackle buildings, with no ventilation and few provisions for personal cleanliness, were constructed to meet war-time orders (Nelson 1975: 141).

Another important factor influencing workers' voice demands during this period was the active, and historically unprecedented, role of government in labor–management relations. The federal government's position on the demands of less-skilled workers for a voice in employment conditions waffled between support for independent unionization and support for internal forms of employee representation, variously labeled "shop committees" or "works councils." When the National War Labor Board (NWLB) was established in 1918, it maintained the following policy through the remainder of the war: workers would have the right to organize trade unions without the interference of employers, but management would be compelled to bargain only with shop committees of the firm's workers, not with independent union representatives.

Thus, during the war the NWLB instituted a policy of requiring employers to grant back pay to any worker who was found to have been discharged due to union activity. In some cases the government went even further, forcing firms to accede to workers' specific demands, but without the accompanying union structure. Workers in meat packing, for example, were granted the eight-hour day with ten hours of pay and a Federal Administrator to arbitrate issues left unsettled by negotiations. In steel, workers were granted time-and-a-half for work over eight hours and for Sundays and holidays, as well as the elimination of a system of bonus pay widely held by workers to be responsible for a general speedup in the pace of work (Brody 1965). In most cases, however, firms with "unstable" labor relations were strongly encouraged to set up internal plans of employee representation and to hammer things out peacefully. Stronger measures were generally applied only when this approach failed.

The NWLB's interest in shop committees was purely instrumental—they would allow parties to an industrial dispute to discuss the sources of disagreement and perhaps bargain over the outcome or agree to some method of arbitration so as to avert a strike. Shop committees were to act as a forum for expressing concerns over wages and hours as well as shopfloor conditions. Clearly, the NWLB had in mind something more than a mere exchange of information between labor and management. As a way of enforcing its desire, the NWLB often oversaw the election of worker representatives in plants, recommended the membership structure of joint labor–management governing boards, and acted as occasional arbitrator in disputes (Montgomery 1987: 415).

An important consequence of this government activity during the war was that workers' views of the illegitimacy of managerial authority and of the injustice in the distribution of shopfloor rewards from production were given "official" sanction. This boosted workers' efforts to organize collectively for institutional change.[15] Steering these efforts to establish a worker voice mechanism in the direction of independent unionization was the American Federation of Labor (AFL), which made a belated, and less-than-enthusiastic, commitment to organizing the less-skilled industrial workers into existing trade union structures. Numerous organizational alliances of skilled and less-skilled workers emerged during the war period. In the metal trades, for example, local councils of workers encompassing a broad spectrum of skills became quite powerful. They were most effective in winning demands in the electrical machine tool, automobile, and shipbuilding industries. These councils served as organizations for the formulation of collective demands, and as the leading wing of the strikes of 1917–18 (Montgomery 1979: 98).

Union membership doubled over the six-year period 1915–20, much of it composed of less-skilled workers drawn into conventional trade union structures. The NWLB's policies and the AFL's new-found commitment to organizing the unorganized were unarguably important for the dramatic gains in union membership over these years. However, the demand for union organization originated from below—from rank-and-file workers. It was fueled, at least in part, by shopfloor discontent, and, given the decline in the exit option, by both the increased desire for and decreased impediments to the formation of a collective voice mechanism which could properly address workers' shopfloor concerns.

Suggestive evidence on the importance of this dynamic in explaining union growth over the period can be gleaned from an examination of the variation in "union growth" across six manufacturing industries between 1910 and 1920 and its relationship to the differing costs of exit to workers as revealed by the industry-level measure of the "reduced turnover" success associated with the implementation of welfare benefits.[16] The simple correlation between "union growth" and "reduced turnover" was statistically insignificant, but further exploration of this relationship yielded an interesting finding.

As part of its survey of welfare activities in 1916–17, the BLS asked establishments whether their welfare programs were jointly administered by management and labor, or by management alone. Using this survey information, it was possible to create a measure capturing the degree of "joint administration" of welfare programs by industry. In industries where joint administration was most common, the relationship between "reduced turnover" and "union growth" should be dampened because a form of worker voice—albeit a limited one—was already in place in a larger percentage of establishments in these industries. Multiple regression results which allowed "union growth" to be correlated with both "reduced turnover" and its interaction with "joint administration" confirmed this dampening effect. More importantly, having accounted for this dampening effect, the relationship between "union growth" and "reduced turnover" became positive and significant at the .10 level. Union growth was more prominent during this period in industries where labor turnover had been most successfully reduced, and unions flourished in these industries if employer-supported mechanisms for worker voice were absent.

Not all mechanisms for worker shopfloor voice which emerged during this period took the form of either government-mandated shop committees or independent unions. Some progressive employers took the initiative to form employee representation plans on their own accord, typically out of a commitment to the principle of worker voice, but also with an eye to the probable consequences of not providing it. In both his book manuscript on the subject and in his writings for the company newspaper—the *Wingfoot Clan*—Paul Litchfield of the Goodyear Tire & Rubber Company expressed the view that:

> The main cause of industrial unrest is the ill-will of the laboring force . . . Any real solution . . . must obtain the good-will and confidence of labor, and this can only be done by direct representation of labor in management.
>
> (Litchfield 1919: 72–3)

The "industrial problem," argued Litchfield, stemmed from workers "who now are very suspicious and trying to overthrow the industries in which they have no voice."[17] The solution lay in granting workers "industrial citizenship"—a voice in managerial affairs. The Goodyear Industrial Assembly was initiated in March of 1919 when Litchfield proposed that a committee composed of appointed management representatives and elected worker representatives study the idea of forming an employee representation plan. The plan was put to a popular vote on June 16, and over 90 percent of eligible voters, almost all of whom voted, favored initiation of the plan.[18]

Employer-initiated plans of employee representation such as the Goodyear Industrial Assembly emerged at a number of establishments during this period. The movement received its start with the Leitch plan for "industrial democracy" which appeared in a few medium-sized plants in the years 1913 and 1914, and then with the Rockefeller plan which was installed in 1915 at the Colorado Fuel and Iron Co. as the result of a commissioned study to recommend a solution to the bitter relations between management and workers in the wake of the Ludlow massacre (Gitelman 1988).

Indeed, it was the employer-initiated form of worker voice—the true company union—that reigned victorious among the various alternatives vying to serve as shopfloor worker voice mechanisms during this period. Neither the NWLB-initiated shop-committees movement or the spurt in independent union growth withstood the test of time. The end of the war-time regulation of industrial relations, the AFL's ultimate inability to overcome its commitment to union structures premised on the skilled trades, and the depression of the early 1920s set the stage for an employer offensive against empowering forms of collective worker voice.

The NWLB made more than 125 awards providing for the installation of "shop committees" (Burton 1926: 29). These were either disbanded altogether or underwent significant changes during the immediate postwar period. At Bethlehem Steel, for example, five days after the armistice, President Grace declared the NWLB's initiatives, including the mandated shop committee, null and void. Grace immediately embarked on the creation of a plan of employee representation with different structural features and, most importantly, without government interference in negotiations and public arbitration of industrial disputes (Montgomery 1987: 414–15).

Independent unions were also eliminated. Union membership reached a peak of 5,047,800 in 1920—or roughly 20 percent of the nonagricultural labor force—but fell rather dramatically over the next three years to 3,622,000 in 1923 (Bernstein 1960: 84). This was accomplished in part through a successful open-shop drive. Employers manipulated postwar patriotism in their attempt to convince the public that the open shop was as American as apple pie, "with equal opportunity for all and special privileges for none" (quoted in Bernstein 1960: 147). However, given employers' determination to remain union free, an open shop was effectively a nonunion shop. Various other measures were employed to ensure the realization of this goal. The early 1920s witnessed increased use of the labor injunction against collective worker action, the "yellow-dog" contract (which required as a condition of employment that workers sign an agreement not to join or organize a union), the blacklist (distributed by growing numbers of employers' associations at the local, industry, and national level), industrial espionage to ferret out union sympathizers, private police forces to ensure the peace, and discharges of union sympathizers. By 1930 union membership as a percent of the nonagricultural labor force had fallen to 10 percent, roughly half the level of only a decade earlier (Bernstein 1960: 84).

Given the success with which employers rid themselves of shop committees and independent unions, it is surprising that they bothered to devote resources to the creation of company-dominated unions. Their central motivation was arguably the realization that while welfare benefits served to reduce voluntary separations, they were much less successful in tackling the problem of worker morale. James Myers, Executive Secretary of the Board of Operatives at the Dutchess Bleachery, Inc., expressed the view of the more progressive employers of the period in arguing that "welfare work" as a solution to "the labor problem" had "been thoroughly tried and found wanting" (Myers 1924: 16). He went on:

> the basic principle of government in industry is the principle of autocracy. All authority is legally vested in the owners ... It is a self-evident

proposition that . . . [among other things] the safety and hygiene of the conditions under which he works . . . are matters of the most vital concern to every industrial worker. Yet he has no recognized voice, is entitled to no direct vote in the control of these matters.

(Myers 1924: 32–3)

Raising the cost of exits for workers had little effect on their perception that managerial authority was illegitimate or that the rewards from shopfloor production were being unjustly skewed towards productivity as opposed to safety and a tolerable work pace.

The war period had served to reinforce workers' ethical condemnation of unrestrained managerial prerogative in production. Through their actions, government officials and union leaders lent credibility and legitimacy to workers' moral and political convictions. The future path of shopfloor relations was affected by the war for primarily this reason. Employers could henceforth strive to alter the form of worker voice, but could not dismiss lightly workers' expressed desire for a say in shopfloor conditions. Doing so would come at a high cost in terms of worker morale and shopfloor productivity.

THE PERIOD OF WORKER VOICE

Compared to just a decade earlier, the mechanism workers used for expressing shopfloor discontent had changed dramatically by the mid-1920s in many large manufacturing firms. Exits had been substantially reduced, and voice—in the form of the company union—had taken their place. Company unions led to improvements in both shopfloor conditions and productivity. They did this, in part, by facilitating an exchange of information concerning workers' preferences for the quality of their work environments and by providing a forum in which management could elicit ideas from workers on mutually beneficial alterations in shopfloor organization. To the extent company unions promoted an exchange of views concerning the legitimacy of managerial authority and the just distribution of shopfloor rewards, they also encouraged greater labor–management cooperation and, thereby, a further enhancement of both parties' shopfloor interests.

That the primary institutional mechanism for worker expressions of shopfloor discontent had in fact changed from exit to voice by the mid-1920s is suggested by the statistics on labor turnover, and especially those on voluntary quits. The average monthly separation rate in manufacturing fell from 10.1 percent over the period 1910–18 to 4.9 percent in the years 1920–9, while the quit rate alone declined by half—from 7.4 percent to 3.7 percent (Jacoby 1983: 268). Turnover rates for the two periods are not strictly comparable, however; the measurements for the latter period are

32

median rates, whereas those for the former period are based on average rates.[19] A comparison of quit rates in the early 1920s with the later 1920s is therefore less fraught with possible error. Monthly quits averaged 5.4 per 100 employees for the period 1919–23 and 2.7 for the period 1924–9 (Lazonick 1990: 251).

The decline in exit behavior was due, at least in part, to the continued growth of welfare benefits and to changing employment practices, including company unions themselves (Owen 1995).[20] A BLS survey of welfare benefits in 1926 revealed significant growth in such benefits since the earlier survey of 1916–17 (US Department of Labor 1917; 1928). Growth was especially strong among those industries offering significant welfare measures in the earlier period, and for benefits tied to length of service such as paid vacations, group insurance plans, and disability benefits. The simple correlation between the "welfare index" constructed from the earlier survey and a new welfare index, capturing the percentage of establishments in each industry offering disability benefits and group insurance plans and based on the 1926 survey, was positive and statistically significant at the .01 level. The industry leaders in the earlier welfare movement continued to be the leaders in the new welfare measures of the late 1910s and early 1920s.

Company unions were central to the changing employment practices of this period. From roughly half a dozen companies employing this form of "employee representation" prior to World War I, the movement grew in fits and starts before reaching maturity around the mid-1920s. Statistics suggest that a stable set of firms with a long-run commitment to maintaining company unions did not emerge until the middle of the decade. Thus, a sizeable number of the company unions that had emerged in the immediate postwar years as simple anti-union measures were gone by the mid-1920s (Nelson 1982a).

Between 1922 and 1924, the number of companies with company unions increased from 385 to 421, but the addition resulted from a gain of 173 companies with newly established plans and the loss of 137 companies that abandoned company-union plans during the period. The number of companies abandoning employee representation plans relative to the average number of companies possessing such plans over the period was 34 percent (National Industrial Conference Board 1925: 10). In contrast, between 1924 and 1926 the number of companies abandoning and adopting plans of employee representation was only 48 and 59 respectively (National Industrial Conference Board 1933: 16). Thus, the number of companies abandoning employee representation plans was only 11 percent of the average number of companies possessing such plans over the period.

By 1926, 913 company unions existed in 432 companies, encompassing roughly 1,370,000 workers (Nelson 1982a: 338), 80 percent of whom were in manufacturing (Nelson 1982a: 344). Company unions existed predominantly among the large, progressive mass-production manufacturers. Their

33

profile greatly exceeded their reach. The labor force of the US stood at roughly thirty million in 1926, and the manufacturing labor force was roughly ten million (US Bureau of the Census 1975: 137). Thus, the percent of the nonmanagerial workforce in manufacturing involved in company unions could not have exceeded 15 percent at mid-decade.

The number of employees covered by representation plans increased continuously throughout the 1920s. This is even true towards the end of the decade when the number of companies possessing representation plans witnessed a modest decline. For example, the number of companies with company unions declined from 432 to 399 between 1926 and 1928, but the number of employees covered by such plans rose from 1.37 million to 1.55 million. The average number of employees per company possessing representation plans also grew over the decade: from 1,792 in 1922 to 3,879 in 1928 (National Industrial Conference Board 1933: 16). This suggests that smaller firms were more likely to abandon plans of employee representation, larger firms were more likely to adopt such plans, and the stable set of firms committed to company unions over the long haul were growing in employment.

Company unions were, at their core, plant-level organizations representing workers' interests to management. Even in the mid-1920s there was enormous diversity in both the form and extent of worker influence through company unions. Typically, though, worker-interest representation was accomplished through elected worker representatives meeting jointly with management representatives for discussion and information sharing. The typical form of organization for worker representation looked something like the following: a plant was divided into districts for purposes of electing employee representatives after a plan of employee representation had been requested (generally, by management). Management representatives (designated as those possessing the power to "hire and fire") were prohibited from either voting or serving as employee representatives.

There were, in addition, age, length of service, and sometimes even English language or citizenship requirements for voting and serving as an employee representative. The ratio of representatives to eligible voters might range anywhere from one to twenty-five to one to every three hundred workers (Burton 1926: 117). Committees of employee representatives would meet, in joint councils, with an equal number of management representatives to discuss matters of mutual concern. Employee representatives also acted as shopfloor-level contacts for worker grievances.

Joint councils had largely advisory capacity, the plant manager being given final authority over decisions. However, in some plans there existed provisions for arbitration when disputes arose. Burton reports, for example, that roughly 30 percent of a sample of surveyed representation plans contained provisions for arbitration of disputes (Burton 1926: 169). However, in only a little over half of these cases was arbitration automatic;

the others required the mutual consent of workers and management (Burton 1926: 172–3). Evidence suggests, moreover, that arbitration was a rarely used device for settling disputes (Burton 1926: 174).

In most respects, company unions of the 1920s bore little resemblance to independent unions. Contact between company unions across plants (even within a single firm) was extremely rare. Almost without exception, company unions did not engage in collective bargaining over wages, hours, or other conditions of employment, and generally had little impact on wages or hours at all. While company unions gave workers a voice over shopfloor issues such as plant safety and sanitation, the power granted to workers was largely that of persuasion. However, when workers' interests squared with those of employers—as, for example, they often did over the capricious activities of foremen or the high rate of workplace accidents—this voice could be powerfully persuasive.

There is plenty of anecdotal evidence to suggest that workers benefited from the existence of company unions. There are a number of recorded instances in which workers clearly initiated representation plans (French 1923: 54). Commons and his staff reported a favorable attitude towards company unions among workers in some of the plants they visited (Commons *et al.* 1935: 348). Leiserson's observations of the company-union movement during the 1920s led him to the following conclusion:

> I think, if you take it as a whole, the unskilled and semi-skilled working people of this country, in the last six years, have obtained more of the things trade unions want out of employee representation plans than they have out of the organized labor movement. Not that they would not have gotten them out of the labor organizations if the labor organizations were efficient in handling the problems of the craftless workers in the mass-production industries.
>
> (Leiserson 1928: 127)

Bernstein states "many workers beyond the pale of the labor movement regarded the company union with its inadequacies as preferable to nothing" (Bernstein 1960: 173).

Various accounts of the activities of company unions offer the strong suggestion that such bodies took seriously workers' shopfloor complaints. Minutes of the meetings of the Industrial Assembly at Goodyear—which were regularly reported in the company newspaper, the *Wingfoot Clan*, during the 1920s—are illustrative. From discussions of ventilation and sanitation problems, more productive and yet safer methods of stopping and starting equipment, procedures for handling machinery during emergencies, and wet and slippery floors, to bolder initiatives, such as jointly sharing layoff decisions with foremen, and joint determination with management of a more elaborate system of job evaluation, the Goodyear Industrial Assembly clearly provided a forum for workers to express their collective discontents, and for management to listen.[21]

One of the responsibilities of company unions was to resolve workers' grievances. Slichter's findings on the resolution of grievances in several large manufacturing firms during this period also suggest that company unions were by no means unresponsive tools of management. For example, of the 2,316 grievance cases brought by company unions at Bethlehem Steel over a five-year period in the 1920s, over 65 percent were settled in favor of the workers. At Swift and Co. prior to 1925, roughly 70 percent were resolved in favor of employees. At GE's Lynn plant, 274 grievances were raised over a two-year period during the 1920s, and 76 were resolved in favor of the workers (Slichter 1929: 413).

Even though company unions were granted very little substantive decision-making authority with respect to working conditions, they offered workers a line of collective communication with upper-level management through joint council meetings of the employee representation plan or through the workings of newly created or expanded personnel departments.[22] Consistent with the notion that voice was a superior mechanism for preference revelation compared to exits, there are a number of recorded instances in which company unions used this collective communication to alter the overall composition of worker benefits—alterations that individual labor turnover was apparently unable to signal to management.

Brandes reports, for example, that through the machinery of the Rockefeller plan at Colorado Fuel and Iron, workers expressed a desire to reduce the workday from twelve to eight hours instead of following US Steel's practice of paying time-and-a-half for the last four hours of the dreaded twelve-hour day. Management acceded to the workers' wishes. Although the change meant a cut in pay, it also brought workers home four hours earlier and set the future standard for the industry (Brandes 1970: 130).

Direct communication with upper-level management also allowed information concerning the shopfloor behavior of foremen to be more widely disseminated. At Swift & Company, for example, the actions of foremen were subject to review by joint councils of the company union during the early 1920s. Management even went so far as to justify the existence of the company union to stockholders on precisely these terms: "The Assembly Plan has also had the effect of making foremen more careful and liberal in their actions and decisions as they come in contact with the workmen from day to day" (quoted in Cohen 1990: 173). Personnel departments systematically encroached on the unrestrained power of the foreman during these years, using the company union as a source of information about the state of shopfloor labor–management relations (Brandes 1970).

Employers appear to have benefited from the existence of company unions as well. Some employers, to be sure, saw their value solely in terms of union prevention. This was especially true in the period immediately following the war. As one employer put it:

> After all what difference does it make whether one plant has a shop committee, a works council, a Leitch Plan . . . or whatever else it may be called. . . . They can all be called company unions and they all mean the one big fundamental point—the open shop.
>
> (cited in French 1923: 35)

However, employee representation plans were by no means a guarantee of worker acquiescence. Nelson maintains that because company unions encouraged collective action among workers, they "probably increased the volume of protest" (Nelson 1982a: 353). A similar conclusion was reached in an internal report solicited by the president of a company considering the establishment of a company union. After "exhaustive research" of existing employee representation plans, the committee of managers preparing the report concluded:

> Even the most successful plans have not stood the acid test. Radical labor agitators have gained entrance into the smoothest running organizations where plant councils have been used for several years. . . . No plan of industrial representation may therefore be construed as complete protection against labor troubles.
>
> (quoted in Weakly 1923: 166)

For many of America's leading corporations, however, company unions brought benefits quite apart from union prevention. After all, employers were in a position by the early 1920s to prevent independent union organization by coercive means; indeed, some did. The company union of the mid-1920s served other purposes.

The company union was an essential component in the peaceful reduction of labor turnover because it acted as a substitute for the exit option as a mechanism for expressions of shopfloor discontent. And reduced turnover enhanced the labor productivity of firms in a variety of ways—for example, in the amount of time they devoted to training new workers.[23] Company unions provided management with workers' insights into the enhancement of productivity. And company unions increased worker morale and fostered greater cooperation between labor and management to the extent they addressed the legitimacy of managerial authority and the just distribution of shopfloor outcomes.

The cost to employers of maintaining an active company union was not insignificant. The Goodyear Industrial Assembly, for example, was one of the more active company unions of the 1920s. It addressed a broad array of worker concerns, including wages, hours, and fringe benefits in addition to shopfloor conditions, and continued throughout the decade to receive the support of both workers and management. Representatives to the Assembly were estimated to have logged an average of 2,860 paid hours of committee work per month during 1925, or nearly one quarter of each representative's workday. Indeed, Litchfield moved in 1926 to limit the company's payment for Assembly work to 1,500 hours per month, and to require that at least three-quarters of every representative's paycheck henceforth be for "productive" work.[24] The action resulted in a two-week "strike" by Assembly representatives, who refused to meet until Litchfield changed his mind. In early March, Litchfield agreed to consider adjusting the hours of paid Assembly work if the new plan could be shown to be inadequate.[25]

Statistics on labor productivity and industrial accidents provide suggestive evidence that worker voice produced benefits for both labor and management alike. Labor productivity in manufacturing grew at an annual average rate of 5.6 percent between 1919 and 1929, compared with a rate of 1.2 percent for the period 1909–19 (Lazonick 1990: 241). Capital accumulation was extensive in the 1920s, and was no doubt responsible for a significant portion of the productivity growth, but many scholars have argued convincingly that labor–management relations also played an important role in the productivity increases during this period (Lazonick 1990; Gordon *et al.* 1982).

A comparison of industrial accident statistics for the period of the 1910s with those of the 1920s is not possible since the Bureau of Labor Statistics did not begin to gather detailed information on accidents until the 1920s. The BLS did, however, conduct a study of industrial safety in the iron and steel industry (reportedly the most dangerous industry during this period) beginning in 1910 (US Department of Labor 1929). The study revealed that the injury frequency rate (injuries per million employee hours) declined from 74.7 to 19.7 from 1910 to 1927.[26] It was in fact the view of many scholars of this period that company unions were an essential ingredient in achieving reductions in accidents. Bernstein cited the area of safety and health as perhaps the "most notable" gain for workers provided by company unions (Bernstein 1960: 172). Safety committees and the grievance procedure allowed information crucial to the attainment of increased shopfloor safety to flow from workers to management.

More compelling evidence on the relationship between company unions and improvements in productivity and safety can be gleaned from industry-level analyses. A combination of data sources allowed the creation of an index of company-union concentration in eight different manufacturing industries for the year 1923 and the accompanying changes in productivity rates and injury rates for a sample of firms in each of those industries over the period 1921–5. Two types of analyses were applied to the data; the results of both appear in Table 1.1.[27]

The first of these employed regression analysis to explore the relationship between company-union concentration and trends in productivity and injury rates. For our sample of industries, productivity growth was 14 percent over

Table 1.1 Suggestive evidence on the relationship between company unions and trends in productivity and injury rates in the early 1920s (sign and significance reported)

(i)Regression analysis

Dependent variables	Company union$_{1923}$	Injuries	Horsepower/ worker$_{1921-25}$	Company union • Horsepower/ worker	N
			Independent variables		
(1) Productivity 1921–25	+***				8
(2) Productivity 1921–25	–	–**			8
(3) Injuries 1921–25	–**				8
(4) Injuries 1921–25	–**		–*	+*	8

(ii) Nonparametic analysis of association

Ranking criterion	Text.	Lumb.	Furn.	Paper	Petro.	Leath.	Elect.	Rub.	Kendall's tau
Company union rank$_{1923}$	6	3	8	5	2	7	4	1	+*
Cooperative shopfloor outcome rank$_{1921-25}$	8	4	7	3	6	5	2	1	

* significant at the 0.10 level (one-tailed test)
** significant at the 0.05 level (one-tailed test)
*** significant at the 0.01 level (one-tailed test)

Key to Abbreviations: Text. = Textiles; Lumb. = Lumber; Furn. = Furniture; Petro. = Petroleum; Leath. = Leather; Elect. = Electrical; Rub. = Rubber.

the period 1921–5, while injury rates declined by 25 percent. The results presented in rows 1 and 3 reveal a strong positive relationship between company-union concentration and changes in productivity rates, and a strong negative relationship between company-union concentration and injury rate changes, as predicted by our institutional analysis of company unions.

Although the small sample size prevents the introduction of many control variables, in rows 2 and 4 we explore the extent to which these results are robust to limited changes in the specifications of the productivity and injury rate equations. The change in industry injury rates is added to the productivity change equation in row 2 in an effort to explore further the company-union effect on productivity. Company unions are associated with improvements in both injury rates and productivity, and yet declining injury rates might be expected to have an impact on changes in labor productivity independent of the company-union effect. The results in row 1 therefore represent a reduced-form estimate of the company-union impact on productivity, but one containing potential bias if injury rates have an independent effect on productivity growth.

Improvements in injury rates can enhance productive performance if fewer labor resources are thereby devoted to training replacement workers or if downtime in production is thereby reduced. Alternatively, safety devices on equipment and worker involvement in safety training programs which enhance safety might come at the expense of labor productivity. The results presented in row 2 suggest that declining injury rates lead to enhanced productivity growth independent of the impact of company unions. More importantly, though, the results also suggest that company unions enhance productive performance entirely through their impact on injuries. The concentration of company unions in an industry had no significant impact on productivity growth except indirectly through reduced injury rates.

The causal relationship running from improved safety to higher labor productivity might be simple and direct—fewer accidents mean less downtime in production and fewer replacement workers who are unfamiliar with their job tasks, and therefore increased labor productivity. The causal relationship might also be more circuitous—safer working conditions improve workers' sense of justice in the distribution of shopfloor rewards, which in turn leads to increased cooperation with management in production and improved labor productivity. Whatever the primary causal explanation, however, the results suggest that there existed untapped opportunities for mutually advantageous changes in shopfloor conditions prior to the rise of company-union voice, and company unions proceeded to tap these opportunities upon their emergence.

In row 4, we present the results of a further exploration of the company-union impact on injury rates. Industries with higher concentrations of company unions were arguably those with a greater devotion to innovations of all kinds, including technological changes in production. Mechanization was continuous and quite rapid over the decade of the 1920s (Jerome 1934), and unreported work examining the simple correlation between company-union concentration and changes in horsepower per worker over the decade confirms our suspicion of a positive association across industries on these two measures of innovation. To the extent that mechanization affects injury rates, the results presented in row 3 may be biased.[28]

The impact of mechanization on injury rates is not straightforward. In the short run, mechanization probably makes the workplace more dangerous as workers attempt to meet production standards while working with new and unfamiliar equipment. In the long run, the effect on injuries could be either positive or negative; mechanization is typically associated with a reduction in physical exertion, which should reduce injuries, but it also allows the speed of production to rise, which probably increases workplace hazards. Thus, it is difficult to say how the introduction of a variable capturing industry-level differences in the pace of mechanization will affect the results presented in row 3.

Whatever the impact of mechanization on injuries, we hypothesize that the relationship between company-union concentration and injury rates will differ depending on the pace of mechanization, suggesting that an interactive specification is appropriate. In particular, we suspect that company unions had a more difficult time garnering improvements in shopfloor safety in an environment of rapid mechanization because workers' knowledge of the equipment and its dangers was inadequate. The ability of company unions to improve workplace safety should be significantly greater among industries where the pace of mechanization was the lowest. The results of row 4 confirm this hypothesis. The estimated coefficients (presented in Appendix Table 1.1) reveal that for average levels of mechanization, company unions contributed to a significant reduction in injuries across industries. A 1 percent increase in company-union concentration produced a 4 percent reduction in the industry injury rate.

The regression results lend support to the view that company unions benefited both management, in the form of enhanced productivity, and workers, in the form of reduced injuries. We can subject this result to a second kind of statistical test—a nonparametric analysis of industry rankings on the two criteria "company-union concentration" and "cooperative shopfloor outcomes" (i.e., the extent of joint improvements in productivity and safety).[29, 30] The results of this analysis, which appear at the bottom of Table 1.1, reveal a positive and significant relationship between the two rankings, lending further suggestive evidence to the claim that the institutional change from exit to voice was indeed mutually beneficial for workers and employers during the early 1920s.

Surveys of the company-union movement often point to important structural changes in the workings of employee representation plans over the decade of the 1920s, leading naturally to the question of whether company unions continued to have a positive impact on shopfloor outcomes into the latter part of the decade. A report by the National Industrial Conference Board cites the following as among the significant structural developments in company unions during this period:

> grievances came to be settled more and more at the source. The employee representative would get his disgruntled constituents together with the foreman or whoever could settle the matter, and in most cases the trouble could be adjusted promptly and satisfactorily.
> (National Industrial Conference Board 1933: 13)

Tendencies towards the decentralization of grievance handling were found in many surveyed firms during the decade. The NICB report found that in one successful works council in textiles, 85 percent of the grievances were settled directly on the shopfloor (National Industrial Conference Board 1933: 22).

The devolution of decision-making and grievance-handling functions, characteristic of company-union structures during this period, was also taking place in management structures, as personnel managers increasingly abdicated recently won decision-making responsibilities in shopfloor management to those foremen and supervisors who had formerly possessed them. This decentralization in decision making was part of a new efficiency-based approach to personnel administration that emerged within the personnel-management movement around the mid-1920s. It emphasized the use of systematic research into industrial relations practices and a largely advisory capacity for personnel administrators as regards production decisions (Spates 1937).

In a retrospective piece on the personnel-management movement during the 1920s, Edward Cowdrick, a leading member of the American Management Association and executive secretary of the Special Conference Committee,[31] contrasted the "euphoric" days of the late 1910s with the more mature, "professional" movement that emerged shortly after the depression of the early 1920s (Cowdrick 1931). Personnel managers would no longer be critics of management, but instead would use professional research findings to advise and assist production departments on the best management techniques. The aim was for enhanced productive efficiency within the broader goal of a more rational treatment of employees. Productive efficiency required that the production department be responsible for managing production, and that authority in production once again rest with foremen and lower-level supervisors (Dietz 1927; Niesz and Knapp 1927).

Did the evolution of employee representation plans and personnel-management structures during this period affect the ability of company unions to influence shopfloor outcomes? Table 1.2 offers suggestive evidence on the

Table 1.2   Suggestive evidence on the relationship between company unions and trends in productivity and injury rates in the late 1920s (sign and significance reported)

(i) Regression analysis

Dependent variables	Independent variables				N
	Company union$_{1924}$	Injuries 1925–28	Horsepower/ worker$_{1925-29}$	Company union • Horsepower/ worker	
(1) Productivity 1925–29	+*				12
(2) Productivity 1925–29	+*	–**			12
(3) Injuries$_{1925-28}$	–				12
(4) Injuries$_{1925-28}$	–*		–	+	12

(ii) Nonparametric analysis of association

Ranking criterion	Industry and rank												Kendall's tau
	C.	Fo.	E.	A.	L.	Pe.	S.	Fu.	R.	Fe.	Fou.	Pa.	
Company union rank$_{1924}$	11	10	3	7	6	2	4	12	1	8	9	5	+*
Cooperative shopfloor outcome rank$_{1925-29}$	12	8	1	7	11	10	3	4	2	9	6	5	

* significant at the 0.10 level (one-tailed test)
**significant at the 0.05 level (one-tailed test)

Key to Abbreviations:
C. = Chemical	A. = Autos	S. =Shipbuilding	Fe. = Fertilizer
Fo. = Food	L. = Leather	Fu. = Furniture	Fou.=Foundaries
E. = Electrical	Pe. = Petroleum	R. = Rubber	Pa. = Paper

relationship between the concentration of company unions and trends in industry productivity and injury rates in the late 1920s. The rate of improvement in injury rates during this period—a 14 percent decline—was less impressive than that of the early 1920s, while the rate of productivity growth was greater, at 23 percent. The analysis presented in Table 1.2 replicates the specifications of Table 1.1 but utilizes data from the latter half of the decade.[32]

The results presented in row 1 suggest that company unions continued to positively affect productivity growth during this period. Comparing the estimated coefficients in the first row of Appendix Tables 1.1 and 1.2, the impact of company unions on productivity growth during the later years was roughly one-half that of the earlier period.[33] The results presented in row 2 of Table 1.2, which control for changes in industry injury rates, suggest that company unions had a positive and significant impact on productivity growth in the late 1920s independent of their influence on productivity through improvements in injury rates. This contrasts with the results for the early 1920s, which suggest that the impact of company unions on productivity growth worked entirely through their ability to reduce injury rates. This independent effect of company unions on productivity growth is consistent with the efficiency-enhancing changes taking place in company union and management structures during this period, which ostensibly fostered greater communication between shopfloor labor and management concerning productivity improvements, but may have also shifted the distribution of shopfloor rewards from workers to firms.

The results presented in row 3 reveal that while company-union concentration is negatively associated with the change in injury rates over the period, the simple association is not statistically significant. This contrasts sharply with the highly significant impact of company unions on shopfloor safety in our analysis of company-union activity during the first half of the 1920s. However, before concluding that company unions lost their ability to influence shopfloor safety entirely during the late 1920s, we consider the possible bias resulting from the omission of mechanization effects in the specification of row 3.[34, 35]

The results of row 4 offer limited support for the view that company unions continued to improve shopfloor safety in the latter half of the decade. Looking at the estimated coefficients presented in Appendix Table 1.2, the results suggest that company unions maintained their ability to reduce workplace injuries where mechanization forces were weak, but where mechanization was strong, company unions were virtually helpless in reducing shopfloor accidents. On average, however, the impact of the company union on injuries was negative and significant. Comparing the estimated coefficients from row 4 of Appendix Tables 1.1 and 1.2, company unions appear to have been roughly one-third less effective in improving shopfloor safety during the latter half of the decade, quite possibly owing to the structural changes in employment practices taking place during this period.[36]

In the bottom of Table 1.2 we replicate the nonparametric analysis of industry rankings on the two criteria "company-union concentration" and "cooperative shopfloor outcomes" for the sample of industries with data on company-union concentration and productivity and injury rate trends for the late 1920s. Although the regression results suggest that company unions became less effective on both fronts during the late 1920s, the nonparametric analysis of association reveals that those industries ranking higher on the index of company-union concentration nonetheless continued to rank higher on the index of cooperative shopfloor outcomes. The mutual advantage of company unions to workers and employers continued into the latter half of the decade.

For the decade as a whole, then, our results suggest that company unions produced improvements in shopfloor safety and productivity. There exists no reliable series on industry injury rates for the entire decade which would allow a further check on our separate findings from the two periods. However, a final check on the impact of company unions on productivity growth can be gleaned from an analysis of industry-level total factor productivity growth over the decade, measures of which do exist.[37] The simple correlation between the "company-union index" for 1924 and "total factor productivity" growth for the decade of the 1920s for a set of eight industries was positive and statistically significant at the .10 level, giving further confirmation to our findings from the earlier analyses.[38]

CONCLUSION

During the first few decades of the twentieth century, many of the large manufacturing firms instituted changes in their approach to shopfloor governance that granted workers a voice in the determination of certain shopfloor conditions. The institutional change from "exit" to "voice" was an historically efficient one; it improved the shopfloor outcomes of both workers and employers through increased shopfloor safety and improved shopfloor productivity compared to establishments that failed to make the change.

While company unions were superior, for both labor and management, to relying on exits for the expression of workers' shopfloor discontent, the transition period from exit to voice nonetheless contained intense labor–management conflict as each party strove to shape the new regime to their own benefit, contesting the power that the other could legitimately hold in production and the shopfloor rewards that the other could justly claim under the new institutional arrangements. This contestation took the form primarily of a struggle over the existence of independent unions.

Company unions won out in this struggle for two reasons: (1) they prevented the distributional losses, in the form of increased wages and perhaps further improvements in working conditions, that employers would have suffered had independent unions been victorious, and (2) they served to enhance the shopfloor benefits of both labor and management, which made them attractive to workers as well as employers. It is the latter that is typically underemphasized in the contemporary literature on company unions.[39]

Evidence that the transition from exit to voice in shopfloor governance, although conflict ridden, was mutually beneficial for labor and management might be seen by some as evidence that the institutional change was both intentional and inevitable. In fact, though, unintended consequences and historical contingency played important roles in the institutional change from exit to voice. World War I exposed the high costs of worker exits as a mechanism for preference revelation over shopfloor conditions and, through the activities of the National War Labor Board, granted legitimacy to workers' demands for voice as an alternative to exits. The emergence of these voice demands, moreover, was in part the unintended consequence of employers' efforts to reduce labor turnover through welfare benefits, which increased the cost of the exit option for workers and encouraged demands for an alternative mechanism for expressing shopfloor discontent. Historical contingency and unintended consequences would play a prominent role as well in the next great change in the institutional arrangements of shopfloor governance—namely, the transition to an empowered shopfloor voice for workers during the 1930s.

2

THE AMOSKEAG PLAN OF REPRESENTATION

The textile mills of the Amoskeag Manufacturing Company[1] date back to the 1830s, when a group of Boston entrepreneurs purchased the water power for the Merrimack River and a 15,000-acre plot of land across the Amoskeag Falls with the intent of building an industrial city centered around textile manufacturing. The company mills, located in the town of Manchester, New Hampshire, closed in 1936 after struggling for most of the preceding forty years to be competitive with the growing southern textile producers, who possessed cheaper labor, newer technology, and a closer proximity to raw materials. The decline in the company's fortunes became especially precipitous following the depression of the early 1920s and the pivotal strike of 1922. The Amoskeag mills failed to earn a profit in four of the seven remaining years of the decade, and the loss in 1924 alone exceeded the combined profit of the three profitable years (Creamer and Coulter 1971: 204).

The events leading up to the emergence of the Amoskeag company union in 1923 conform quite closely to those of many large manufacturing firms during this period. The context within which the Amoskeag company union operated during the 1920s, however, was atypical in at least one very important respect. Because of the company's dire financial condition, throughout the decade workers faced declining real wages, longer hours, and continuous attempts by management to speed up production. The atmosphere in which labor and management cooperated to enhance productivity and shopfloor safety through the employee representation plan was thus filled with great tension and occasional conflict. For this reason, the Amoskeag company union provides an interesting case study. To the extent that evidence can be marshaled suggesting that, despite such tension-filled and conflict-ridden conditions, the Amoskeag company union constituted a mutually advantageous institutional adjustment in shopfloor governance, the case is strengthened for the positive impact of company unions more broadly.

The nature of workers' shopfloor discontents in the early-twentieth-century textile mills of the Amoskeag varied across departments. Preparatory efforts such as carding and spinning were typically very dirty jobs. The weaving rooms were generally kept clean, but they were hot and stuffy places to work because windows had to remain closed (the wind played havoc with the weaving process, causing threads to break and flying fabric to build up on the rigs). Weaving was also the noisiest of the textile operations. Bleaching and dyeing, which took place after weaving in the case of cotton, but occasionally before weaving in the case of woolens, involved the use of unhealthy chemicals. The risk of injury was probably the greatest in the mechanical and maintenance departments. The general work pace, which had never left much room for worker socializing, became a major source of worker discontent during the mid-to-late 1920s when the company tried to compete with the southern mills by getting more work from the workforce.

Little is known about the extent to which worker discontent is responsible for the high labor turnover at the Amoskeag during the early years of this period. Brissenden and Frankel's research on labor turnover during the 1910s involved a detailed look at accession and separation rates at the Amoskeag in 1914. Their study revealed a separation rate of 123 percent of the labor force for that year (Brissenden and Frankel 1922: 176–7). Company documents suggest a similarly high separation rate for 1913: voluntary separations were 71 percent of the labor force of cotton weavers and 142 percent of the labor force of cotton spinners during the period.[2] These were the two largest occupational categories in Amoskeag's textile mill employment at the time.

The cost to the firm of turnover rates such as these is difficult to document. We do know, however, that a financial reorganization of the company in 1911 was accompanied by a major effort to reduce unit labor costs, and that the primary institutional innovations which were introduced to accomplish this goal were the initiation of an employment office to oversee the hiring and firing of labor and the introduction of welfare measures to reduce labor turnover (Creamer and Coulter 1971: 172). Over the next five years, and with increasing rapidity following the Lawrence textile strike of 1912,[3] the company introduced the usual gamut of welfare measures, from health services (medical and dental) and housing subsidies (purchase and rental) to pensions and profit sharing. Evidence that these welfare measures had an appreciable effect on labor turnover is largely anecdotal. A report by the employment department in the fall of 1915, for example, claimed that departmental activities had produced some success at reducing turnover in the preceding four years (Creamer and Coulter 1971: 203).

Worker demands for a voice in the determination of employment conditions emerged at the Amoskeag around the end of World War I in the form of the organizing activity of the United Textile Workers (UTW). The UTW never achieved official recognition by the company, at least not as judged by the standards of industrial unions in the period after World War II, but it did come to possess a significant presence in labor–management relations at the Amoskeag between the years 1918 and 1922. Real wages rose rather significantly during the period 1918 to 1920, and the dreaded 55-hour week was reduced to 48 hours. These accomplishments were not due solely to union activities. The reduction in work hours, for example, was prompted in part by the decline in production following demobilization of the war effort, and by an expectation that several New England states would, in the very near future, pass legislation requiring a reduction in the length of the work week. But the union's presence was clearly a factor.

For purposes of shopfloor voice, the union set up grievance committees in various departments of the plant, and succeeded in convincing management to establish an adjustment board which met with these committees to adjudicate worker grievances. By most accounts, the union's shopfloor presence was significant in getting management to recognize workers' complaints and in counterbalancing management efforts to increase the pace of production in response to the wage increases of the period. Union-initiated limits on the number of frames or looms to be run by workers were commonly cited in both union and adjustment board meetings during 1920 and 1921 (Creamer and Coulter 1971: 190–1). The union could not officially enforce such limits, of course, but its endorsement of production limits granted a certain legitimacy to workers' shopfloor efforts to resist further speedups in production.

Beginning in early 1921, the company embarked on what appears to have been a systematic effort to alter its already unofficial relationship with the union. Management began, for example, circumventing union grievance procedures by allowing the adjustment board to meet with disgruntled workers outside the bounds of the grievance committee structure (Creamer and Coulter 1971: 192–3). In October of 1921 the union lodged a formal complaint of antiunion activity with the adjustment board, claiming that overseers systematically favored nonunion workers and that the feeling among union members was that management was committed to "breaking up the union organizations" (quoted in Creamer and Coulter 1971: 194).

When, in early 1922, management abruptly announced that wages would be rolled back 20 percent and the work week would be increased to 54 hours a week, a nine-month strike ensued. Concessions from workers were demanded by other New England textile mill owners at this time as well, but not all textile workers in the region had obtained the reduction of hours of the Amoskeag workers, who objected strenuously to the increase of hours, and correctly viewed management's announcement as an attempt to re-establish the compensation package workers had possessed prior to the UTW organizing drive of 1918. Some seven months into the strike several textile manufacturers withdrew their demand for the wage cut and the Amoskeag management did the same. Many Amoskeag workers attempted to stay out in opposition to management's continued demand that work hours be increased, but upon withdrawal of the demand for a wage reduction a back-to-work movement ensued that allowed the company to begin production some two months later.

With the resumption of production, the company eliminated most of its welfare measures, rooted out most of the remnants of union influence, and initiated the introduction of an employee representation plan. To most observers of the company's industrial relations policy during the 1920s, the company union was seen as an institutional device by which the company could introduce further wage cuts and speedups while avoiding labor strife. Is it possible, however, that a plan instituted under such dire circumstances could be attributed with improving shopfloor conditions and enhancing productive efficiency?

The Amoskeag Plan of Representation received its start in May of 1923, when at management's initiation a group of workers, most of whom had been strikebreakers during the tumultuous events of the previous year, got together to discuss the idea of establishing an employee representation plan. Following a favorable review of the idea from this group, management arranged for the nonmanagerial workforce to elect worker delegates, a subset of whom would be given the responsibility for drawing up, in cooperation with management, the structural features of the plan of worker representation. A committee of twelve workers was chosen from the 244 elected worker delegates, and in September a plant-wide referendum was held on the plan they had developed.

The results of this vote are illustrative of the state of labor–management relations at the time, and of the environment within which the Amoskeag company union operated during this early period. The UTW urged its members to vote "no" on the proposal. Workers in the mechanical and maintenance departments, who had no affiliation with the UTW, voted overwhelmingly in favor of the plan. But only four of five departments in the worsted section voted for it, and only one of the five departments in the cotton section voted in support of the plan. In cotton spinning and cotton weaving the votes were nearly two to one in opposition.[4]

Management promptly placed the entire workforce in the cotton section on layoff, blaming "business conditions." After a petition was circulated among workers requesting a re-vote, the remaining worsted department was brought into the plan in October, and the entire cotton section was finally added in January of 1924. Creamer and Coulter's assessment of the events surrounding the plan's inception led them to conclude:

> It is clear from the manner of its inauguration that there was no widespread sentiment for the company union and that it was virtually forced upon the workers.
>
> (Creamer and Coulter 1971: 210)

The structural features of the Amoskeag Plan of Representation were similar to those of many prominent company unions of the period. In fact, the plan was formally modeled after the company union at International Harvester. Worker representatives were chosen in secret ballot elections and

50

were paid for the time they devoted to company-union activities. The formal machinery of the plan was composed of a hierarchy of joint labor–management committees, all containing equal numbers of labor and management representatives. Joint labor–management committees were formed in each department of the plant. Further up the hierarchy, there were three joint sectional committees for the cotton, worsted, and mechanical sections as a whole. And, finally, three "general joint committees" were formed to address the specific issues of "Routine, Procedure, and Elections," "Safety and Health," and "Production Methods and Economy."[5] While management was not compelled to accept any committee decision, unanimous decisions were generally considered settled issues (Creamer and Coulter 1971: 212).

As in many employee representation plans of the period, representatives were urged to refer a grievance first to supervisors for resolution before bringing it before the formal machinery of the plan of representation. Much of the activity of the Amoskeag company union appears to have taken place at this level, making it difficult to assess the overall impact of the plan because lower-level shopfloor labor–management communications were rarely recorded. However, the minutes of various joint committees do exist. Looking over these records from company documents, one is struck both by the amount of time and energy devoted to the consideration of worker grievances and by the wide variety of employment concerns which were raised.

Worker demands for wage adjustments, both in the aggregate and across jobs to reflect differing job content, were regularly considered by the employee representation plan at the Amoskeag during this period. The company union also appears to have been a forum in which management decisions concerning internal labor allocation could be questioned. Thus, numerous examples exist of workers petitioning through the company union to be reinstated following an unfair layoff decision or to be reassigned to a former job after an unfair assignment by an overseer.

Addressing shopfloor issues was a primary concern of the Amoskeag Plan of Representation. The shopfloor condition that provoked the most discontent among workers during the 1920s was probably the pace of production. The worsening financial condition of the company led management to continuously pressure workers through the stretch-out and other forms of speedup resulting from changes in job standards and descriptions. Even on this, however, the Amoskeag Plan of Representation appears to have been somewhat successful at forcing management to take seriously workers' concerns about the intensity with which they worked.

In the spring of 1924, for example, the "box loomfixers" in the cotton weaving department complained of the increased company work standards and requested that work standards be rolled back to their 1921 level without any cut in pay. After much investigation, including apparently the standards and pay at other mills, the joint departmental committee recommended, and the superintendent conceded to, a 7 percent reduction in the number of box looms per section as the new work standard, with no decrease in pay.[6]

Somewhat later in the decade, workers won through the company union a commitment by management to conduct a time study analysis of any job at the request of a joint departmental committee.[7] Beginning in 1926, and with increased rapidity in the later years of the decade, joint departmental committees routinely requested that certain jobs undergo time study. Available evidence does not allow us to say whether the work pace was ever significantly reduced as a result of time study analysis, but by merely possessing such a right the Amoskeag workers stood in a more privileged position than most industrial workers during these years.

Other shopfloor issues appear to have been less contested by management. Creamer and Coulter (1971: 214) conducted a thorough accounting of all grievances that came before joint committees of the plan of representation at the Amoskeag over the nine years of its existence, along with whether they were decided in favor of workers or management. Table 2.1 records the results of their investigation. Employees were victorious in their complaints almost 40 percent of the time. They won over 30 percent of their wage and labor allocation grievances, over a quarter of their speedup grievances, and over 60 percent of their working conditions complaints.

Drinking water availability, ventilation problems, the condition of wash-room facilities, and safety concerns related to moving belts, lighting, and

Table 2.1 Number of docketed complaints, by decision and nature of complaint, at the Amoskeag textile mills [a]

Nature of complaint	Total cases	Number decided	
		For employees	Against employees
Total	200	77	123
Specific wage rates and hours	87	32	55
Discharges, lay-offs, and transfers	49	16	33
Stretch-out	31	8	23
Other working conditions	33	21	12

[a]Allocation of a decision in favor of or against employees was based on the authors' understanding of the complaint and the nature of the ruling. All complaints, 1923–32, are included. From Creamer and Coulter (1971, Table C-1)

machinery operation methods are just some of the issues that appear in the minutes of meetings of the General Joint Committee on Safety and Health.[8] Committee decisions capture only a small part of the impact of the company union on shopfloor conditions, however, as many such issues were raised and resolved at the less formal level of worker representative and overseer on the shopfloor itself.

Fortunately, we need not rely solely on descriptive statistics on the resolution of working conditions grievances, a recitation of the minutes of committee meetings, or speculation about the success of worker representation at the shopfloor level to ascertain the impact of the Amoskeag Plan of Representation on shopfloor conditions. It was possible to construct from the files of the corporation an account of workplace accidents for the period 1920 to 1927, and thereby to assess the accident performance of the company both before and after initiation of the employee representation plan.[9] The number of lost-workday injuries (i.e., injuries requiring the loss of at least one day of work) during this period appear in Table 2.2.[10] Statistics for 1922 are omitted because of the significant lull in productive activity during this year due to the strike.

Because yearly employment figures are not available for the entire period of interest, we take two different approaches to estimating employment for the years 1920 to 1927.[11] The first, which appears in the second column of Table 2.2, is created by interpolation of the employment figures for 1919 and 1929, both of which are known (Creamer and Coulter 1971: 70). The second, appearing in the third column, utilizes employment figures available

Table 2.2 Injuries at the Amoskeag textile mills

Year	Lost-workday injuries	Employment$_1$*	Employment$_2$**	Injury rate$_1$	Injury rate$_2$
1920	342	11,793	11,736	0.030	0.030
1921	424	11,510	11,396	0.037	0.037
1922	–	–	–	–	–
1923	334	10,944	10,716	0.031	0.031
1924	203	10,661	10,376	0.020	0.020
1925	252	10,378	10,034	0.024	0.025
1926	235	10,095	10,843	0.023	0.022
1927	249	9,812	10,670	0.025	0.023

* Employment1 is based on interpolation of employment figures for 1919 and 1929

** Employment2 is based on employment figures for 1925–27 and interpolation of employment figures for 1919 and 1925 from company records for the period 1925 to 1927, and interpolates between the years 1919 and 1925 to arrive at yearly employment estimates for the earlier period. The first estimate yields a smooth trajectory of employment decline over the decade of the 1920s at a rate of 283 workers per year. The second estimate contains a smooth decline in employment during the first half of the period, but fluctuates thereafter, reflecting actual movements in company employment. The different approaches yield very similar results.

The corresponding injury rates (i.e., the number of lost-workday injuries divided by the labor force) appear in the last two columns of Table 2.2. The average injury rate for the period 1920 to 1923—prior to the company union's existence—is .033 regardless of the employment series used, while the average for the period 1924 to 1927—after introduction of the company union—is .023 using the employment figures in column 2 and .0225 using the employment figures in column 3. A statistical test of the difference in mean injury rates for the periods before and after the initiation of the plan of employee representation reveals that the differences are indeed significant at the .01 level, irrespective of which approach is taken to estimate yearly employment figures.

These results offer suggestive evidence that the plan of employee representation acted to reduce accidents in the mills.[12] The reduction in the injury rate is all the more impressive given the severely curtailed hours of work during the depression period of 1920 and 1921, the increased length of the workday beginning in 1923 (from roughly 48 to 54 hours per week), and the unceasing efforts of management in the latter half of the period to speed up production. The evidence would seem to suggest that worker voice produced a one-time improvement in mill safety, but given the increased speed of production in the later years of the period, continuing improvements in safety may well have been required to prevent the workplace injury rate from rising.

Is there any evidence to suggest that the Amoskeag company union also served to enhance productive efficiency? Given the financial circumstances of the corporation during the 1920s, and the pressure placed on workers to increase the pace of production as a consequence, a comparison of productivity growth both before and after the establishment of the employee representation plan would be misleading. Any enhanced growth in labor productivity could arguably be attributed to the speedups of the period, and not to the efforts of the employee representation plan. There is evidence to suggest that workers viewed their acquiescence to the speedups as part of a cooperative effort to enhance the long-run viability of the corporation, and that the Amoskeag company union played a limited roll in fostering this cooperation. But this is hardly the sort of evidence that would compel us to claim that the Amoskeag Plan of Representation actively boosted the productive efficiency of the enterprise.

There is rather compelling evidence to suggest, however, that the company union played a crucial role in preventing management's strategy of remaining competitive through labor speedups from becoming excessive, and thereby threatening productive efficiency. Beginning as early as 1924, joint departmental committee meetings were grappling with worker complaints concerning the impact of speedups on the quality of materials and productive efficiency. Such complaints became even more common during the later years of the decade.

In 1927 the cotton weavers complained that the increased work standard of 32 looms per weaver was counterproductive given "the quality of warps and filling as they are at present" (quoted in Creamer and Coulter 1971: 26). Downtime due to broken threads was excessive when one worker was required to tend 32 looms. The weavers were able to convince management of their claim, and it was decided that the standard should be reduced to 28 looms per weaver until the quality of materials being supplied to the weavers could be increased sufficiently to make the increased work standard feasible.

The inferior quality of intermediate materials was often the result of unproductive work standards in earlier stages of production, as revealed in the complaints of worsted weavers to their joint departmental committee in the winter of 1928.[13] In an effort to economize on the costs of transporting materials from the spinning rooms to the weave rooms, management had increased the size of the filling trucks which were used for such purposes. Weavers noted, however, that the increased standards—which required transporting workers to lift trucks weighing in excess of 100 pounds—resulted in "mixed and oily filling being woven into the cloth," presumably making weaving more difficult and lowering the quality of the finished product.

These, and other examples from company documents, offer suggestive evidence that joint committee meetings of the plan of representation provided a forum in which worker representatives could explain to management the ill-conceived consequences of increased production standards and other desperate attempts to boost productivity and lower costs of production.

CONCLUSION

The adoption of a plan of employee representation at the Amoskeag Manufacturing Company during the early 1920s represented a mutually advantageous institutional change in shopfloor governance for both labor and management alike. Given the competitive pressures facing the firm in the 1920s, both labor and management were clearly better off during the 1910s, before the rise of the company union. The evidence suggests, however, that labor and management would have been even worse off had the employee representation plan not existed. It is in this sense that worker voice was mutually beneficial for labor and management. Injury rates declined following institution of the employee representation plan. And, while most other aspects of employee compensation worsened during the 1920s, the Amoskeag company union sought to insure that managerial decisions concerning the pace of production both possessed some semblance of fairness, through time study analysis for example, and translated into real productivity gains.

3

THE RISE OF AN EMPOWERED
SHOPFLOOR VOICE

Although workers in the more progressive manufacturing firms had been granted a voice in the determination of certain shopfloor conditions during the 1920s, shopfloor discontent had by no means disappeared from the industrial scene. Liberal labor policies, including company unions, were confined in their reach to a subset of the manufacturing base; even such notables as the Ford Motor Company and US Steel did not possess company union representation for their workers on the eve of the depression. While the cost to workers of exiting had risen significantly in many of these firms, in some cases due to the maintenance of high wages, and in others due to the sluggish growth in employment opportunities elsewhere, plant-level mechanisms for the expression of worker voice remained nonexistent.

Moreover, even among the more progressive firms that possessed company unions, worker voice was limited in scope to a subset of workers' most pressing employment concerns. Wages and hours were rarely the subject of discussion in company-union meetings to be sure, but even with respect to shopfloor conditions—the terrain within which company unions had achieved their greatest success—there were important limitations. By most accounts of the period, mechanization and wage incentive schemes placed increased pressure on the pace of production, an issue company unions had difficulty regulating to the mutual advantage of labor and management. And the decentralization of shopfloor decision making during the late 1920s meant that the company union's influence in the area of the behavior of foremen had been severely circumscribed. The 1930s depression exposed the limited reach and scope of the company-union movement of the 1920s.

SHOPFLOOR SETBACKS DURING THE EARLY DEPRESSION YEARS

During the early years of the depression, aspects of the drive system of production returned to many industries that had abandoned this approach in the 1920s. Speedups were the most common worker complaint. Rubber workers in Akron, Ohio, for example, complained bitterly of the increased pace of production during this period, commenting "They just run the life out of a man" and "No time to eat—or go to the wash room—unless you lose production" (Nelson 1988: 114–15). Bernstein (1970: 99) claimed that the speedup was the "most obnoxious" of the rubber workers' grievances during the early depression period.

A similar story existed in textiles. "Stretch-out" was "the constant cry" (quoted in Bernstein 1970: 299). "When you get out (at the end of the day), you're just trembling all over," offered one textile worker. "Tired, tired and weary—like all the others," said another (quoted in Bernstein 1970: 299).

Shopfloor conditions also worsened during the early depression years in industries where the drive system had not been firmly challenged by the personnel-management movement of the 1920s. The pace of production increased in the auto industry, for example, where workers had been complaining for decades about the speed at which they were driven to work. One auto worker wrote to President Roosevelt that "no human on earth can keep up with the lines," adding "the way they are using these moving lines sure is slavery" (quoted in Fine 1963: 14). A report by the National Recovery Board on depression conditions in the automobile industry concluded, "Depression competition has spurred the speed-up beyond economic capability to produce day by day" (US Department of Labor 1935a: 646).

The pace of production had been a source of worker discontent in industries both within and without company unions before the onset of depression, but the downturn in business activity made matters worse. The policy, promulgated by the Hoover Administration, of having employers forgo wage decreases during the early days of the depression led many firms to pressure foremen to step up the work pace in hopes of stemming the tide of declining profits. When, in the fall of 1931, this policy of holding the line on wages could no longer be maintained, wage cuts in many industries, especially where piece rates or bonus payment schemes were used, led workers themselves to increase output per hour of work in an effort to counter the deterioration in living standards. Foremen continued to pressure workers wherever possible to speed up production.

The reduced length of the workday in many manufacturing concerns during the depression years also may have contributed to the increased pace of production. Tire manufacturers, following Paul Litchfield's lead at Goodyear, reduced the workday to six hours in the early 1930s in an effort to share the work among their respective workforces. While for Litchfield, and perhaps for others as well, this was largely a humanitarian gesture, Litchfield noted that the change might also allow productivity to increase "if we were in a position to push production" on a less exhausted and hungrier labor force (Nelson 1988: 114). Apparently, the experiment worked. Per unit labor costs were reportedly cut by more than 8 percent at the India Tire Co. with the reduction to six hours per day. The superintendent of production

found that workers on six-hour shifts did not have to rest or pace themselves, while the need to replace lost income made them willing to take on
more production (Nelson 1988: 114).

Significant improvements in labor productivity were surprisingly
common among industries during the depression. The rubber tire industry
witnessed a rapid improvement in labor productivity between 1929 and
1933 (Nelson 1988: 114). Steel output per worker hour was larger by a
third in 1937 compared to its level in 1929. And in the packing houses
during 1934, 40,000 workers were doing the work that 52,000 had done in
1929 (Cohen 1990: 317). Labor-saving technological change, improvements
in machinery, and alterations in the composition of products were partly
responsible for these increases in labor productivity, but increased labor
intensity on unaltered tasks and the increased pace which generally accompanied increased mechanization also played an important role.

Another common complaint of workers during the early depression years
was the treatment they received at the hands of foremen.[1] Workers' discontent with the behavior of foremen was partly related to the latter's visible
role in the increased pace of production during the period. But there were
also basic human indignities workers suffered under the direction of
foremen, indignities that appeared to be exacerbated by the depressed
economic conditions.

Foremen could affect the pace of work through a variety of methods. In
addition to closer worker supervision or adjusting the speed of automated
assembly operations, foremen were able to vary the tasks associated with a
particular job. While a select number of firms had begun developing
systems of centralized job descriptions and job evaluation schemes during
the period of the personnel-management movement, most firms in the early
1930s left the specific details of a worker's job to foremen, and regulated the
pay attached to jobs by establishing a range of hourly or bonus rates for a
broad class of job categories. The pressure placed on foremen during the
depression to increase productivity, in combination with their power to alter
the content of jobs, promoted speedup through an expansion of worker
responsibilities. The stretch-out in textiles is a prime example.

Another method by which foremen affected the pace of work during the
early depression years was through their ability to influence decisions
concerning layoffs and job assignments. The personnel-management movement had made important inroads during the 1920s into the foreman's
ability to hire and fire indiscriminately, but the depression left firms with a
gargantuan task of deciding who would continue to work and at what kinds
of jobs, the basis for judgment being generally some combination of worker
merit, ability, seniority, and need.[2] The first two of these criteria required
significant input from foremen. And for many foremen, the workers with
the most "merit" and the greatest "ability" were simply those that worked
the hardest. Thus, to be kept on, or to be assured of the better jobs in the

plant in the event of one's job being eliminated, an intense commitment to production was often important. As one autoworker put it, "if you were one of the bastards who worked like hell and turned out more than production—you might be picked to work a few weeks longer than the next guy" (Fine 1969: 61).

In addition to the workers' discontent with foremen based on their influence over the pace of work, there was an increased psychological burden workers suffered during this period as a result of the enhanced dictatorial tendencies the depression appeared to foster in foremen and lower-level supervisors. Verbal abuse, and worse, accompanied the foreman's directives. Indeed, workers appear to have resented these indignities more than the physical burden of the increased pace of production. As an auto worker complained to Senator La Follette in the midst of the Flint, Michigan, sit-down strike, "We were treated like a bunch of dogs in the shop and we resented it so much that the people with principle . . . were grabbing for anything to try to establish themselves as men with a little dignity" (Fine 1969: 59). The National Recovery Board report on depression conditions in the auto industry concluded that "Labor unrest exists to a degree higher than warranted by the depression." In addition to the problem of speedups, the report noted that "The foreman's power and the gap between the workers and the executive are important causes of labor unrest" (US Department of Labor 1935a: 646).

Such treatment was not limited to industries, such as autos, where the progressive reforms of the 1920s had failed to make inroads. Art McCollough, a skilled motor assembler who had been demoted to the refrigerator assembly line at the General Electric plant in Erie, Pennsylvania, recalled the sense of injustice and powerlessness he felt in the face of abusive behavior by foremen: "Why you could stand there and hear a foreman bawl a guy out for something that didn't amount to anything. Make him look like a damn fool. And you couldn't do nothing about it, see" (quoted in Schatz 1983: 62). General Electric was at the forefront of the welfare capitalist movement of the preceding decade.

It is often claimed that the foundation of the welfare-capitalism movement crumbled during the early depression years, thereby ushering in the need for the government to act as an alternative provider of welfare benefits, and independent unions to act as an alternative worker voice mechanism. In fact though, the basic structure of the employment reforms of the 1910s and 1920s remained largely intact during the early depression (e.g., Spates 1937). In March of 1934, a survey of 233 major manufacturing companies revealed that with few exceptions—such as dismissal compensation and special bonuses—employment procedures in place in 1929 were maintained during the early depression years (National Industrial Conference Board 1934: 10). For instance, roughly 75 percent of reporting concerns possessed centralized procedures for hiring and firing, and less than 2 percent of these concerns abandoned such procedures during the depression.

Some welfare benefits were indeed eliminated during the early depression years, but the record is one of general resiliency. Stock purchase plans virtually disappeared, and profit sharing and paid vacations for wage earners were significantly cut back, as were many recreational activities (National Industrial Conference Board 1934: 4). But less than 2 percent of reporting firms eliminated their group life insurance plans; less than 3 percent eliminated mutual benefit association programs such as accident, sickness, and death benefits; and pension plans were eliminated in less than 7 percent of the reporting concerns.[3]

A census of company-union activity over the period reveals that the number of companies with company unions fell by 22 percent between 1928 and 1932, from 399 to 313; the number of separate organizations (i.e., establishments) maintaining company unions fell by 12 percent, from 869 to 767; and the number of employees covered fell by 18 percent, from roughly 1,500,000 to 1,250,000 (National Industrial Conference Board 1933: 16). However, compared to the 21 percent decline in the number of manufacturing establishments (US Department of Commerce 1935: 718) and the roughly 40 percent decline in manufacturing employment (US Department of Labor 1935b: 22), these numbers indicate a probable *increase* in the concentration of company-union activity among the manufacturing workforce during the early depression years.

Worker discontent in the more progressive firms during this period was due less to the deterioration of past reforms than to a growing awareness of their significant limitations. The depression exposed the holes in the structure of 1920s employment practices without fundamentally jeopardizing the structure itself. For instance, although 75 percent of the sample of 233 major manufacturing firms cited above possessed centralized control over hiring and firing prior to the depression, less than a third maintained systematic promotion plans or rating schemes that could have been used in place of the recommendations of foremen to make layoff decisions (National Industrial Conference Board 1934: 10). Formal procedures for determining layoffs were not abandoned during the depression years, they were simply not in place to a large extent prior to the depression.

While welfare programs remained largely intact, the greatest welfare needs of workers during the early depression period were for employment security or unemployment compensation, neither of which had been formal commitments of employers during the earlier reform period. Indeed, despite their absence during the earlier period, numerous employers made exemplary efforts at addressing the new needs of workers in this area during the depression. Work-sharing efforts were fairly widespread among progressive firms. Wisconsin Steel, General Electric, and International Harvester offered loan programs to employees during the depression. And Western Electric,

American Rolling Mill, and Standard Oil of New Jersey granted termination allowances to those long-term employees who had to be permanently let go (Brody 1980: 69; Cohen 1990: 243–6).

As for company unions, we know very little about their activities during the early depression period. However, despite whatever impact they continued to exert on such things as industrial safety, shopfloor cleanliness, and health-related matters, company unions had never been very effective in addressing those issues that were of major concern to workers during the early depression years. Few company unions appear to have made much headway in influencing the pace of production. Time studies conducted in the presence of company-union representatives, and discussions of their results in joint council meetings of representation plans, were the exception rather than the rule. Early on, when personnel managers held more power in production activities, company unions may have had some influence in eliminating the most egregious behavior of "driving" foremen, but this ability deteriorated during the late 1920s because of decentralization of shopfloor decision making and then again during the early depression period.

Thus, on the eve of the National Industrial Recovery Act (NIRA) of 1933, workers' shopfloor discontent was on the rise due to the increased pace of production and the inhuman treatment they were receiving at the hands of foremen. And yet open expressions of discontent, in the form of strikes and union organizing activity, in the previously unorganized sectors of manufacturing, are hard to come by. Workers at the US Rubber plant in Mishawaka, Indiana, spontaneously walked out in the spring of 1931 to protest the company's introduction of a new task and bonus system and the inability of an elected committee of employees to resolve the shopfloor discontents this system created. The workers complained of the impossible pace required to meet the new production standards, of having to start work an hour or two early in order to make an efficiency rating that would keep them on the payroll, and of pay falling by more than a third for much more work (Brody 1980: 74–5).[4] In general, though, workers seemed cowered by fear and tired by hunger. The NIRA and the slight upturn in economic activity in 1933 changed things dramatically.

THE SEARCH FOR AN EMPOWERED SHOPFLOOR VOICE

The NIRA period was like a throwback to the late 1910s in terms of the incentives it provided workers in their efforts to build a collective shopfloor voice. Shopfloor discontent among workers was high, due, in this instance, to the competitive pressures placed on firms by the depression and the consequent deterioration in shopfloor conditions. The cost of expressing shopfloor discontent through worker exits was significant, owing, in this case, to the high rates of unemployment. And the government was once again actively supporting a collective voice for workers in determining their conditions of employment, this time because of the social turmoil created by depression, rather than the need for industrial peace during time of war.

However, a crucial difference between the two periods is that the potential cost to workers of building an empowered voice mechanism was much higher during the early depression years. Slack labor markets fueled competition between workers for scarce jobs, and made the ultimate price one could pay for organizing activity—i.e., loss of a job—much greater than during periods of tight labor markets such as existed during World War I.

Several features of the NIRA, in combination with the mild upturn in economic activity during the second half of 1933, served to reduce the costs of worker organizing, thereby leading to increased demands for an empowered shopfloor voice. Two organizational forms upon which workers might have grafted an empowered shopfloor voice mechanism existed during the early depression period: reformed company-union structures and labor unions chartered by the American Federation of Labor. Workers in many industries experimented with both; neither proved viable in the final analysis.

The NIRA was signed into law on June 16, 1933, in an attempt to restore recovery to a crisis-ridden economy.[5] It proposed to regulate competition between producers, and between employers and employees, through codes of fair competition; to redistribute purchasing power among consumers through various government relief efforts; and to focus the "united action of labor and management" towards the goal of recovery. Section 7(a), which gave workers "the right to organize and bargain collectively through representatives of their own choosing . . . free from the interference, restraint or coercion of employers . . . in the designation of such representatives," applied to all employers covered by codes. It was thought to be an essential ingredient in fostering the "united action of labor and management" towards the lofty goal of economic recovery.

Section 7(a) was an ideological shot in the arm for workers with significant shopfloor discontents. It granted a certain legitimacy to workers who spoke out on the need for a collective worker voice in production decisions, and it altered the consciousness of workers who before had either failed to perceive such a need or were reticent to act on it. Beyond that, however, there were problems of both conception and enforcement.

The Act left open the exact nature of the voice mechanism to which the government would give its support. Section 7(a) clearly prevented employers from forcing workers to join company unions as a condition of employment, but it did not make them illegal if workers chose them as forms for worker representation. It did not spell out the process by which worker representatives would be chosen—whether, for instance, there would be elections and, if so, whether the majority would rule or there would be proportional representation. The Act did not elaborate on the appropriate unit for worker representation, thereby leaving open a question of importance to the debate surrounding craft versus industrial unionism. And enforcement of the right to organize unions in the absence of employer coercion remained inadequate throughout the Act's existence.

Clarification of the government's position on many of the organizational issues would come in future labor board decisions and in the 1935 National Labor Relations Act (NLRA). Following enactment of 7(a), a series of labor boards were instituted to clarify the statute's intent with respect to worker representation. President Roosevelt created the National Labor Board (NLB) in August of 1933, and special labor boards were created to handle cases arising in particular industries. An early version of the National Labor Relations Board (NLRB) was then created in June of 1934, by congressional resolution. It was given the right to investigate controversies, hold elections and hearings, and render decisions on violations of 7(a). The Supreme Court held that the NIRA was an unconstitutional delegation of congressional power in its 1935 Schechter decision—not because of Section 7(a), but because of the codes regulating business affairs—thus giving rise to the need for the NLRA in July of 1935.[6]

Arguably the most significant of the pre-NLRA decisions was the NLRB's Houde decision of August 30, 1934.[7] It established the doctrine of representation elections, majority rule, and the proper subjects of bargaining. In its discussion of the proper subjects of bargaining, the language of the decision contains a consciously ill-concealed swipe at the activities of 1920s company unions:

> Toilet facilities, safety measures, lighting and ventilation, coat racks, slippery stairs, and so on . . . in no sense constitute the recognized subjects of collective bargaining, namely, wages, hours and basic working conditions.
>
> (National Labor Relations Board 1934: 38)

The Houde decision also stated very clearly that the purpose of collective bargaining was the negotiation of a written collective-bargaining agreement which would stabilize the terms of employment for both parties for a fixed duration.

There were other statutory features of the NIRA period, in addition to Section 7(a), that had a substantive impact on workers' demands for an empowered voice mechanism to resolve shopfloor grievances. One was the government's effort to provide relief for the unemployed through unemployment benefits and public employment. The activities of the Federal Emergency Relief Administration and the Civil Works Administration served to reduce one of the most important costs a worker might face for engaging in union organizing activity—the loss of employment.

Estimates of the coverage and extent of benefits associated with these early relief efforts suggest that their impact was not insignificant. For instance, roughly 60 percent of the 10.5 million people estimated to be unemployed during April of 1934, were receiving public relief (Lescohier 1935: 256). Moreover, wages in the newly created public sector jobs were initially comparable to, or even better than, those prevailing in the private sector (Lescohier 1935; Peterson 1935). While the average private sector wage for unskilled labor had fallen from forty cents an hour to less than half this amount during the preceding years of depression, the minimum Civil Works Administration rates were forty cents in southern states, forty-five cents in the middle states, and fifty cents in northern states (Lescohier 1935: 255). By the end of 1934, it became clear that the government could not employ the throngs of workers willing to work at such generous wages. Accordingly, local work relief administrations were instructed in November of that year to pay the wage prevailing in the local community.

Unemployment benefits through the Federal Emergency Relief Administration were not as generous. However, the agency did maintain a policy of granting relief to the needy irrespective of whether or not they also happened to be on strike. It is difficult to say just how large a role this played in the growing strike activity beginning during the summer of 1933, but the policy gained a national reputation during a number of strikes of this period, most notably the 1934 textile strike. During the strike, the Illinois Manufacturers Association expressed the opinion that the policy was an invitation to "universal promotion of industrial warfare" by "organized minorities" (quoted in Bernstein 1970: 308).

Another feature of the NIRA period that arguably influenced workers' demands for a greater voice in determining shopfloor conditions was the regulatory codes governing working conditions in specific industries. These codes offered workers government confirmation that many of their particular shopfloor discontents were indeed legitimate. Working conditions codes covered minimum standards for safety and health, including lighting, ventilation, machine guards, and protective clothing (e.g., US Department of Labor 1934: 1,089–93), but they were also occasionally known to address such heated topics as work loads. The government maintained an insufficient staff and lacked certain statutory powers to enforce the codes, but as workers became aware of them, or felt their impact in firms where codes were initially adhered to only to be abandoned later on, the codes served as a legitimating force for workers' open expressions of shopfloor discontent.

There was a tremendous surge in strike activity beginning almost immediately after the passage of the NIRA in June of 1933. Worker-days lost to strikes, which had not exceeded 603,000 in any month during the first half of 1933, jumped to 1,375,000 in July and to 2,378,000 in August. Due largely to the rise in strike activity during the second half of the year, the number of strikes in 1933 exceeded that of any year since 1921 (Bernstein 1970: 172–3). The strikes were predominantly for union recognition, and a great many of them took place in industries that possessed a history of union experience, such as coal, clothing, and textiles.

One of the most famous of the NIRA period strikes in manufacturing was the nationwide walk-out of textile workers in the fall of 1934 (Bernstein 1970: 298–315; Gerstle 1989: 127). Organizing activity in textiles had resumed following the passage of the NIRA after more than three decades of rocky performance by the United Textile Workers of America. Although employers were vehemently opposed to unions, the upturn in business activity convinced textile firms to abide by NIRA industry codes which limited production and thereby reduced the glut of textile goods on the market, but which also contained regulations on wages, hours, and working conditions, including work-load restrictions. When the mild recovery stalled in the summer of 1934, employers began to disregard the codes, and work loads increased. Wildcat strikes emerged during the summer of 1934, and the nationwide strike began in September as an outgrowth of workers' disgust with employer unwillingness to abide by government codes.

The strikes of this period were not entirely confined to industries with established union histories; there were rumblings in such places as rubber, steel, and autos as well. In August of 1933 workers at the Auto-Lite company of Toledo, Ohio, formed a federal labor union, chartered by the AFL, covering the auto parts manufacturer and related operations in Toledo. When the company refused to recognize the union (on one occasion ordering its business agent off the company premises), the workers went on strike. The radical American Workers Party became deeply involved in the strike at this point, instituting mass picketing and rallying the community— employed and unemployed alike—behind the cause. The oppressive tactics of the local authorities, which involved a combination of injunctions and police brutality, and the willingness of the Auto-Lite company to employ strikebreakers, led to a bloody confrontation know as the "Battle of Toledo" (Bernstein 1970: 222). When the smoke cleared, two men had been killed, fifteen wounded by gunfire, and ten guardsmen required medical attention.

Mechanisms for collective worker voice spread rapidly following passage of the NIRA. Company unions led the pack. Employers with little company-union experience responded to the rumblings of workers over deteriorating shopfloor conditions by instituting plans of employee representation. An April 1935 survey by the Bureau of Labor Statistics revealed that roughly two-thirds of the establishments possessing company unions at the time, and encompassing nearly 60 percent of the workers represented by company unions, had initiated them during the period following the NIRA (US Department of Labor 1938b: 50–1). In total, only 5.2 percent of the manufacturing establishments reported possessing company unions, but this represented over 33 percent of the sampled manufacturing workforce and exceeded the 28 percent that possessed trade union representation (US Department of Labor 1938b: 35,37).

The numbers were much more impressive among the major manufacturers. A March 1934 survey of 233 large manufacturing companies found that 54 percent of the concerns maintained employee representation plans, roughly a third of which had been adopted since the NIRA (National Industrial Conference Board 1934: 8). In the fall of 1933, Goodrich, Firestone, and General Tire created company unions in their tire plants, adding to those already existing at Goodyear and US Rubber (Nelson 1988: 125). The auto industry, which possessed virtually no plans of employee representation prior to this period, quickly adopted company unions; by 1934, practically all of the firms in the industry possessed one (Fine 1963: 155). In steel the number of employee representation plans grew from seven prior to the NIRA to ninety-three by the end of 1934, and the percentage of steel workers affected rose from 20 percent to over 90 percent (Brooks 1940: 79).

There was no disguising the employer motivation for initiating such plans. One steel operator commented, "We put in an employee representation plan because we were afraid our men would fall for this outside union stuff which we were sure would come in right after the passage of the Recovery Act" (Brooks 1940: 78). Survey evidence confirms that this was indeed the dominant motivation. A detailed study of ninety-six company unions begun during the NIRA period revealed that in sixty-three of the cases company unions were initiated by management in reaction either to a recent strike or to the fact that a trade union was making some headway among workers in the plant or the locality. In thirty-one cases, management cited the NIRA as a dominant concern in its introduction of employee representation plans, and in only two cases did management report that a desire to improve personnel relations led them to the introduction of a company union (US Department of Labor 1938b: 81).

The strategy among many of these employers was to do battle with "outside unions" by showing workers that the company union could prove superior as a worker voice mechanism. T. G. Graham, the Vice-President at Goodrich, argued, "I believe that industry with the right viewpoint can train men more rapidly for Employee Representation work than can the A.F. of L." (quoted in Nelson 1988: 129). By March of 1934, the Goodrich plan boasted the institution of a minimum daily wage, an appeals policy for discharged workers, and the resolution of many worker grievances (Nelson 1988: 130). These accomplishments—like those of most company unions during this period, and unlike those of most company unions of the 1920s—were produced in the presence of significant outside union agitation, making them difficult to attribute solely to the plan of employee representation.

Many employers were pleasantly surprised with their initial company-union experience, and lamented the years of mutually advantageous gains their companies had lost due to the absence of employee representation plans. A steel executive, looking back on the experience with company unions in 1933–4, commented:

But just because the employee representation plans were put in to protect us from outside unions, don't make the mistake of thinking that that's their only value to us now. We've gotten a real education out of our experience. We hadn't realized, as we grew bigger, how far away we had drifted from personal contact with the men in the plant. We ... knew nothing of the men's feelings and grievances. The foremen would often be hardboiled with them and pay no attention to their complaints. . . .

(quoted in Brooks 1940: 78–9)

As company unions expanded in these firms, so too did the personnel-management practices characteristic of the best practice employers of the 1920s (Jacoby 1985).

Most of these plans began in the same way as they had in the early 1920s, with employers initiating them and developing the structure. However, the combined impact of 7(a), emerging labor board interpretations, and workers' demands for empowerment led to structures of employee voice that were much more favorable to labor than those of the preceding decade. For instance, workers were typically granted greater independence from management in the new company unions. The joint (labor and management) committee arrangement, so common to the company unions of the 1920s, was replaced by a structure in which employees met separately from management in order to formulate demands. Roughly two-thirds of the NIRA-period company unions maintained this structural independence (US Department of Labor 1938b: 133). Workers were also able to raise a broader set of concerns with management than they had been able to raise in the past—including not just health and safety issues, for example, but also work pace, wages, and hours—and did so in a fashion much more resembling bargaining than discussion. In addition, there was greater contact between company unions across the various plants of a particular company (US Department of Labor 1938b: 72–3).

The experience of company unionism at US Steel is illustrative. Shortly after the emergence of plans for employee representation, workers began making bolder demands in discussions with management: wage increases, more liberal pensions, paid vacations, surrender of the unrestricted right to fire by management, and arbitration of grievances. When, at one point, local management of a Carnegie-Illinois subsidiary replied that such decisions had to be made at the corporate level, the company union made contact with other company unions, culminating in the demand upon the company to recognize the consolidated organization as a multi-plant bargaining agent (Bernstein 1970: 456; Brooks 1940: 83–5).

Company-union structures that pre-date the NIRA period began to mimic the newer arrivals in structure and purpose. Company unions in meat packing, for example, were given greater substantive powers, including "negotiating" with employers over wage rates and paid vacations (Brody 1964: 155). Company unions at GE and Westinghouse were restructured and granted informal bargaining rights; management representatives were removed, the restructured voice mechanisms were renamed "workers' councils," and management-initiated informal negotiations with workers' elected council leaders (Schatz 1983: 66). The evolution of company-union structure and conduct continued into the period following the NLRA, when the company-dominated union had become, by law, an unfair labor practice.

The National Labor Relations Act was enacted on June 5, 1935, less than six weeks after the NIRA was found unconstitutional by the Supreme Court in its Schechter decision. Section 7 of the Act expanded on and clarified the rights of workers contained in section 7(a) of the NIRA. It encouraged "the practice of collective bargaining" and protected "the exercise by workers of full freedom of association, self-organization, and designation of representatives of their own choosing, for the purpose of negotiating the terms and conditions of their employment or other mutual aid or protection." Section 8 allowed for implementation of these rights by prohibiting certain unfair labor practices of employers, one of which, under subdivision (2), is "to dominate or interfere with the formation or administration of any labor organization or contribute financial or other support to it."[8]

While not outlawing company unions per se, the Act defined a "labor organization" very broadly, thereby restricting employers' influence over virtually any worker voice mechanism that purported to deal with management concerning any condition of employment. The meaning of the phrase "to dominate or interfere with" was not properly spelled out in the Act itself, but its meaning emerged more clearly from various NLRB decisions in the years immediately following the Act's passage. Plans that do not allow for employees to meet independently of management; plans where the choice of matters to be discussed is controlled by management; or plans where the election of representatives is arranged, conducted, and supervised by management were generally found to be in clear violation of section 8, subdivision (2). Forms of worker voice had to be made consistent with the workers' freedom of association, as guaranteed by the Act, and old-style company unions could not pass this test (Brody 1994).

Employers' reactions to the NLRA varied. In some instances, company unions underwent even further evolutionary transformations, as employers attempted to address some of the objections to existing employee representation plans found in emerging NLRB decisions. In other cases, employers openly maintained plans of representation that were in clear violation of the law. Regardless of their responses, employers were unanimous in their hope that the Act would be found unconstitutional. They felt the Act to be a restraint on local relations between employer and employee and local conditions of production, neither of which the federal government could legitimately regulate because neither was associated with interstate commerce.

In *Matter of International Harvester Co.*, the NLRB left little doubt about its fundamental objections to company unions, even in the form they had arrived at by the mid-1930s.[9] The NLRB concluded that because the employer typically controls the purse strings of such organizations (and certainly did so in the case of the Harvester Industrial Council), employees are effectively denied the help of outside experts who possess information necessary for the intelligent discussion of complex problems involving the fixing of wages, hours, and conditions of employment. Lacking independent financial support, employees represented by company unions are also left with no recourse when a deadlock is reached between labor and management. Finally, the employees in such plans typically lack a binding collective labor agreement with their employer. "As a result," stated the Board, "its (International Harvester's) employees possess only the shadow, not the substance, of collective bargaining."

Those conditions deemed necessary by the Board to ensure that employees possess the "substance" as opposed to merely "the shadow" of collective bargaining went to the very heart of employers' fundamental opposition to independent unions—namely, that they were influenced by "outside" forces and that they desired to negotiate as equals with management in drawing up a legally binding labor contract.

After the Supreme Court rendered its decision on the NLRA in April of 1937, declaring, in the case of Jones and Laughlin, that the right of employees to self-organization was essential to industrial peace, and that the latter was essential to the free flow of goods and services through interstate commerce, employers were forced to deal with the new reality. Where outside unions had not already forced their abandonment, employee representation plans either disappeared or evolved into independent local unions.

In meat packing, the plans took on such names as the Packing House Workers' Union of Sioux City, Iowa (at Cudahy) (Brody 1964: 170). In the electrical sector, the "workers' councils" at GE and Westinghouse took on titles like the General Electric Independent Union, and started collecting dues and holding membership meetings (Schatz 1983: 67). A 1938 NLRB study of independent local unions found that 93 percent of those responding to the survey had been formed during the period following the NLRA, 65 percent had appeared immediately following the Jones and Laughlin decision, and 92 percent had some connection to a previous company union (cited in Brooks 1939: 70).

Many of these new unions would not withstand the combination of NLRB scrutiny and competition from outside union forces. A number did survive, however, by winning support from a majority of workers in representation elections.[10] Those that eventually succumbed to outside unionization managed, moreover, to fend off such forces for years longer than similarly situated establishments that did not possess independent local unions. Whether those that remained were truly "independent" is difficult to discern.

Company-dominated unions were formally eliminated from the industrial scene in the US by government edict. However, many workers had come to recognize drawbacks in this approach to collective worker voice prior to government intervention. Where employers were particularly cynical in initiating employee representation plans and reluctant to listen to and act on the voices of discontent from their employees, it took little time for workers to see through the pretensions of company unionism. What were workers to think of incidents such as those at Chrysler's Dodge Main plant, where management removed the chairman of the works council after accusing him of making "a slur on the President of the Chrysler Corporation" by suggesting that the company's rhetoric of hard times was a sham because "there were hidden profits" (quoted in Jefferys 1989: 109)? Constructing the vision of a better alternative, rather than convincing workers of the need for one, became the challenge for activists in such firms.

In other cases, however, there was a more serious challenge facing those wishing to foster the emergence of an empowered worker voice. For a significant number of workers, substantive improvements in worker welfare accompanied the various machinations of company unions during this period. Workers in these firms had to be made to see that the accomplishments of company unionism were largely due to the pressure placed on employers by the threat of outside unionization and by the growing militancy of workers' internal activities via informal work groups and aggressive shopfloor leaders.

Some workers came to see this reality on their own, and became early leaders in the outside union movement. Other workers needed convincing. More progressive leaders infiltrated company unions, attempting to expose their weaknesses and to show workers that when concessions were granted by management, this was done because of workers' rising power. The government's evolving position on employer dominance through company unions also helped to convince workers that employee representation plans were voice mechanisms adequate for the expression of collective demands, but not for collective bargaining. Another important force helping to tip the scale away from company unions was the growing political power of workers. Especially with the re-election landslide of Roosevelt in 1936, workers began to see that political democracy could make a difference, and that if democracy in the political arena made sense, it made sense in the economic arena as well. The company-union movement granted workers a voice in certain shopfloor decisions, but it was a far cry from industrial democracy.

In embracing this fact about company unions, contemporary observers often fail to see, however, just how instrumental employee representation plans nonetheless were in helping to establish something closer to industrial democracy. William Leiserson was aware of this potential as early as the late 1920s when he wrote of company unions:

71

Once they are created, human organizations of this kind have a way of evolving according to laws of their own nature in quite unforseen directions . . . .The management has started a movement in the direction of democracy in industry that is bound to grow.

(Leiserson 1928: 122–3)

Company unions contributed to this "movement in the direction of democracy in industry" in several ways. They provided workers with formative experiences in representative democracy and in collective-interest communication, if not negotiation, with employers. Workers received practice in voting and running for office, and elected officials learned skills in collectivizing and prioritizing workers' concerns and in conferring with management. Company-union shopfloor representatives, the analogue to the shop stewards of future local union structures, were encouraged to regularly confer with foremen and supervisors, thereby acquiring valuable communication skills and confidence in dealing with shopfloor management. Representatives to joint council committees, the analogue to local officers in future union structures, acquired similar skills and confidence in communicating with upper-level management.

In industry after industry, local union organizing activity was led by men and women who had served lengthy stints with their employers and had been representatives in their firm's company unions. Schatz's (1983: 81, 85) detailed analysis of the union pioneers in electrical manufacturing is illustrative. Of the fifteen people on the executive board of United Electrical, Radio and Machine Workers of America (UE) Local 506 at the Erie General Electric plant in 1940, all but one had been at the plant since before 1925. Out of a group of twenty-eight union pioneers in the electrical industry for which there exists evidence, at least eight (or roughly one-third), had served as representatives in company-union representation plans.

Certain structural features of company unions also served as models for future union organizations (Schacht 1975). The plant-wide nature of worker representation in company unions contributed to workers' preference for and support of union organization on an industrial basis. Even when districts for worker representation in company unions were drawn along craft lines within departments, as they often were, the commingling of representatives on committees instilled in worker representatives a sense of both the commonality of worker grievances and the need for collective representation of workers' interests in order to ensure success in negotiations with management. In instances where workers' concerns involved aspects of production beyond the control of plant management—such as production standards or the responsibilities of the personnel department, which were often set at company headquarters—the inability to influence such matters sent clear signals to workers concerning the need to organize across plants in order to attain certain goals.

The structure of shopfloor representation in company unions also served as a model for similar representation through independent unions. Jefferys argues, for example, that the formal and informal shopfloor structures of the company union at Chrysler's Dodge Main plant were "a model for the rights of shop stewards" under future union structures and an important force in the development, even before the company's recognition of the union, of a steward-based system of shopfloor organization (1989: 108). Indeed, the steward system evolved within the shell of the company union, with UAW stewards winning forty-seven of the fifty-three districts in the February 1937 election of company-union representatives (Jefferys 1989: 114).

Independent unions of the American Federation of Labor (AFL) were the second organizational form vying for workers' allegiance during the early years of the depression period. In some industries, organizing appeared in the form of federal labor unions. Between June and October of 1933, the AFL and its member unions issued charters to 3,537 federal labor union locals (Millis and Montgomery 1945: 193–8). Over 40,000 rubber workers possessed membership cards in AFL-chartered federal labor unions during the NIRA period, and perhaps as many as 100,000 auto workers were enrolled (Bernstein 1970: 95,100).

In other industries, less-skilled workers swarmed into already-existing AFL union structures. Large numbers of steel workers joined the Amalgamated Association of Iron, Steel and Tin Workers, an AFL union with an industrial charter and a long and famous history dating back to the late nineteenth century (Brooks 1940). In meat packing, where the Amalgamated Meat Cutters and Butcher Workmen of North America had been chartered by the AFL in 1897 (Brody 1964), similar events unfolded during this period. Textiles workers flocked to the United Textile Workers of America, a national union chartered by the AFL decades earlier, and which on the eve of the 1934 general strike could claim a membership of 350,000 (Galenson 1960: 325).[11]

In most cases, however, these early union structures never achieved a recognized presence in labor–management relations, and membership was fleeting. In rubber, for example, membership in the federated AFL locals went from a high of over 40,000 in early 1934, to less than 4,000 by September of 1935. In steel, membership in the Amalgamated had reached 150,000 during 1934, only to plummet to 10,000 a year later. And in autos, after starting out as the primary union presence in the industry, AFL federal locals accounted for less than half of the unionized workers in the industry by 1935 (Bernstein 1970: 372).

The AFL approach to organizing the less-skilled workers suffered from numerous problems, two of the more important of which had to do with a dispute between the organization and those it was attempting to organize over the vision of the union movement for mass-production industries. The AFL viewed the federal locals as temporary organizations that would exist only until workers could be properly doled out to one or more of the various craft unions that supported the organizing campaign (Zieger 1986). The less-skilled workers desired industrial unions not craft unions, however, and thus their support for the AFL's organizing efforts tended to be almost as short-lived as the federal locals themselves.

The second problem facing the AFL's efforts was that the organization lacked a firm commitment to addressing the less-skilled workers' shopfloor discontents, and was unwilling to condone the militant tactics required for workers to achieve an empowered shopfloor presence on their own. The AFL's foremost desire was to build stable union structures from which future gains might be made; it strove to accomplish this end by relying largely on government protection rather than rank-and-file direct action.

This aspect of the AFL's failed approach is nicely revealed by the events at General Tire in the summer of 1934, in one of the first important sit-down strikes to take place in American industry (Nelson 1988: 136–42). The workers' major grievances were the declining piece rates and related speedups in production that made their pay and conditions lag behind those offered by the other major tire manufacturers. When the company union was unable to make any headway with management on the workers' concerns, local labor leaders decided to push for company recognition of the existing AFL Federal Labor Union 18323 and a collective-bargaining agreement to address the workers' grievances.

A representative for the AFL suggested that the appropriate path towards this end was for the local union to seek exclusive representation rights through the regional labor board, and to use the attainment of this official status as a way of forcing General Tire management to recognize and bargain with the union. However, given the early stage of development of labor law and the weak power of labor boards at this time, this approach seemed destined for failure. Local leaders thus took it upon themselves to press their demands through a sit-down strike, and expended enormous effort over the course of several weeks coordinating shopfloor activity to ensure the success of the endeavor.

When on June 19, the local union president demanded and failed to receive a response from the company to workers' demands, he walked through the plant giving the signal for work to stop in the various departments. After a brief period during which workers remained inside the plant, the conflict turned into a more conventional strike lasting over a month in duration. The company finally capitulated with an increase in wages and the elimination of the company's support for the company union, but refused to formally recognize the union local. More significantly, however, was the fact that AFL officials were a party to none of the concluding negotiations. Support for the local's affiliation with the AFL had waned significantly over the course of the struggle.

Later in the decade the Butcher Workmen (BW) of the AFL and the

Packinghouse Workers Organizing Committee (PWOC) of the Congress of Industrial Organizations (CIO) competed for the allegiance of packinghouse workers in a campaign that reveals similar sorts of weaknesses in the AFL's approach (Brody 1964: 173–8). The two organizations committed roughly the same resources to the organizing effort. Moreover, by this time the AFL had adopted a commitment to industrial union organization, so both the AFL and CIO were committed to building similar union structures. The AFL was nonetheless no match for its CIO rival.

In his careful analysis of the emergence of unions in this industry, Brody cites two reasons for the AFL's lack of success: the PWOC maintained a strong commitment to rank-and-file participation in the initial organizing and in the later running of the union itself, and it displayed a willingness to have workers invoke militant tactics, such as the slowdown and quickie strike, to press for the resolution of grievances. These features of the CIO approach, perhaps more than anything else, sealed its fate as the structural form out of which would emerge an empowered shopfloor voice for workers.

THE ATTAINMENT OF AN EMPOWERED SHOPFLOOR VOICE

The emergence of an empowered shopfloor voice for workers in mass-production manufacturing took place during the latter half of the 1930s. It emerged as the product of two central forces: (1) the growing power, militancy, and cohesiveness of informal work groups; and (2) the industrial organizing drives of workers to establish independent unions. Its existence following unionization was premised on an uneasy combination of informal and formal institutional arrangements—shopfloor norms and customs of worker militancy on the one hand, and collective bargaining and grievance procedures on the other.

The actions of informal work groups to limit production and establish worker control over shopfloor conditions dates back centuries in time. The earliest, most famous citing in twentieth-century American manufacturing is associated with the writings of Frederick Taylor, who referred to these activities as worker "soldiering" and put forth principles of "scientific management" to combat them (Taylor 1919). Employers utilized a number of tactics over the first few decades of the twentieth century to limit the power of informal work groups to exert control over production, most of which ultimately proved unsuccessful.

One such tactic was incentive payment schemes, which had been recommended by Taylor in his writings, and which spread rapidly across many manufacturing industries in the 1910s and 1920s. Incentive pay was supposed to increase the value to workers of acting individually rather than collectively in regulating their work effort. However, employers consistently disregarded Taylor's admonition that incentive pay should not be reduced once workers were forthcoming in increasing their work effort. The widespread practice among employers of reducing bonus rates once pace had been significantly enhanced sent a clear signal to workers that noncooperative behavior could lead to a worsening of individual and collective worker welfare, thereby encouraging the efforts of informal work groups to collectively regulate work pace.

Another tactic used by employers to reduce the power of informal work groups in production was to foster ethnic diversity within such groups. Prior to the collective-voice demands of workers in the late 1910s, this was not a popular employer strategy; indeed, it was common for employers to segregate workers by ethnicity (as well as by race and gender) in the assignment of workers to work groups.[12] In the steel mills of South Chicago, for example, Italians worked as unskilled laborers in the bricklaying department, Poles served as unskilled workers in the blast furnaces, and Croatians provided the common labor for the finishing departments of rail mills.

During the 1920s, employers attempted to reduce their vulnerability to acts of worker solidarity based on ethnic affiliation by dispersing workers from the same ethnic group more widely across jobs in the plant. At Wisconsin Steel in the mid-1920s, a mill superintendent made note of the new arrangement to a visiting researcher:

> You see here how it is arranged. We try never to allow two of a nationality to work together if we can help it. Nationalities tend to be clannish and naturally it interferes with the work and the morale of the place. You see here in this loading department, for instance, we have an Italian, an Irishman, a Pole, a Dane and two Mexicans working.
>
> (quoted in Cohen 1990: 165)

Employers fostered ethnic diversity within work groups in the hopes of rendering collective norms for worker output much more difficult to create and enforce. However, this tactic—much like that of incentive pay—proved to be a failure. The strategy relied for its success on members of ethnic groups maintaining a divisive form of identity. But, at the same time that employers were busy dividing up ethnic enclaves in the workplace, they were also engaging in paternalistic efforts to weaken workers' ethnic identity, and the sense of difference that flowed from it.

Company-sponsored pension plans, sickness benefits, and life insurance plans were, at least in part, efforts to redirect workers' loyalties from ethnic communities to the firm, by substituting employer welfare benefits for the mutual benefit activities commonly provided by ethnic communities. Company recreational activities were meant to substitute socializing across ethnic groups of workers for socializing within ethnic groups in the community. Company sports teams in baseball, track, swimming, and bowling combined workers from various ethnic backgrounds to compete against teams from other companies in industrial leagues. Company picnics provided the time for workers of different ethnicities to socialize, and sing-alongs, employing tunes from American popular culture,

76

provided the material that would bind workers together from diverse ethnic backgrounds (Cohen 1990).

The naturalization campaigns and patriotic symbolism of the "Americanization" movement in public policy during World War I were brought within the firm during the 1920s in the form of English language classes and civics lessons.[13] Employers thereby hoped to contribute to the abolition of left political movements and the diversity of political opinion among the immigrant workforce.

Ethnic tension by no means disappeared during the decade of the 1920s, and tensions over race and gender differences were at best only slightly diminished, but employers' efforts to reduce solidarity among workers within ethnic enclaves and to eliminate the diversity of political commitments must be counted as partially successful, especially among the second generation of immigrant workers. However, breaking up ethnic groups in the workplace and breaking down ethnic ties in the community provided the foundation for the increased homogenization of working-class consciousness, an important ingredient in plant-level worker solidarity and the success of union organizing drives in the 1930s.

The pressure workers faced to increase the pace of production also contributed to the building of strong informal work groups. Rapid mechanization during the decade of the 1920s, the decentralization of shopfloor decision making at firms that had earlier experimented with centralized control through personnel-management departments, and the spread of wage incentive schemes placed increased pressure on workers to perform tasks more quickly. Company unions typically made little headway with management when it came to workers' concerns about the pace of production. Thus, the only worker device that could successfully counteract the mounting pressure was the cooperation of informal work groups to limit the pace of work.

In a host of industries, and in a variety of ways, workers appear to have engaged in shopfloor cooperation to regulate output during the 1920s. Mathewson's (1931) often-cited participant-observer study of the restriction of output among unorganized workers during the late 1920s attests to the significance of informal work groups in regulating work pace during this period. Tire builders at Firestone set up a limit on production that was not surpassed "regardless of what the company did" in an effort to regulate the impact of mechanization and piece-rate cuts on the work pace (Nelson 1988: 93). Women workers in the sliced bacon department at Swift adopted methods to deceive rate setters during time-study periods, returning to more efficient procedures after the threat of altered work standards had passed (Cohen 1990: 202). Bank wiremen at the Hawthorne Works of Western Electric maintained a strict code of solidarity in output regulation which forbid "rate-busting" or "squealing" to management about informal shopfloor production norms (Schatz 1983: 44).

Informal work groups became the building blocks for plant-wide cooperative efforts to form unions during the 1930s. (The coordinated work-group effort during the General Tire conflict of 1934—discussed above—is illustrative of developments during this period.) Militant expressions of shopfloor discontent during the decade originated with the activities of informal work groups, and were crucial exercises in union organizing. And even following successful organization, union recognition by the company, and attainment of a collective-bargaining agreement, workers' shopfloor power would come to rest on the strength of informal work groups and shop stewards to regulate shopfloor conditions.

Union organizing activity under the banner of the CIO in the late 1930s was the other instrumental force in the rise of an empowered shopfloor voice for industrial workers. The Committee for Industrial Organization was formed in November of 1935 at the behest of John L. Lewis and with the financial support of the mineworkers, clothing workers, and ladies' garment workers unions. It was initially formed to work within the AFL to "promote organization of the workers in the mass production and unorganized industries" (Bernstein 1970: 401). The next few months witnessed the slow alienation of the Committee from the AFL. Charges of dual unionism were followed by the AFL Executive Council's instruction in September of 1936 for all affiliates to sever their ties with the Committee or face suspension from the Federation. By this time, the affiliates counted among them the rubber, steel, auto, and oil workers in addition to the mine workers, clothing workers, and ladies' garment workers. The various affiliates refused to comply with the Executive Council's order, and the CIO was born.[14]

The CIO had become, by 1937, a national entity staffed by deeply committed organizers capable of directing broad-based organizing drives among scattered industrial plants. However, its distinguishing feature in relation to earlier organizing efforts was its willingness, at least initially, to condone and even encourage rank-and-file participation and militant shopfloor action in the formative stages of unionization.

It was this organizing activity, more than the resulting union organization, that led to the emergence of workers' shopfloor empowerment during this period. In the successful struggle to form a union, workers combined the cooperation of informal work groups with accumulated experience in shopfloor dispute resolution through company unions into a powerful shopfloor presence. The process of organizing and the struggle to win union recognition revealed to workers the power that lay latent in strategic shopfloor cooperative activity. The developmental forces of the earlier period—informal work groups, company unions, and worker homogeneity—provided the necessary background conditions for the emergence of workers' shopfloor power. Its attainment came through a series of actions that altered workers' views of themselves and their collective abilities *vis-à-vis* employers.

The most visible of these actions were the sit-down strikes of the late 1930s. The rubber workers of Akron, Ohio, were the pioneers of this tactic. Five sit-downs occurred at Firestone, Goodyear, and Goodrich between January 28 and February 18 of 1936 (Nelson 1982b: 212), thus inaugurating what would turn out to be a tumultuous year of shopfloor guerilla warfare in rubber. The Firestone struggle was the first to occur, and is illustrative of the kinds of issues that sparked such militant behavior among industrial workers. The company had cut piece rates in the tire room earlier in the month, but the tire builders, all members of the budding United Rubber Workers (URW) local, acted in unison to slow the pace of production instead of competing with one another and increasing output to maintain hourly income. When a nonunion "rate buster" was moved to the tire room to break the pace restriction, he and the local URW committeeman came to blows and the committeeman was suspended for a week. His fellow workers sat down and stopped work in protest.

By year's end auto workers were adopting similar tactics in their struggles to resist managerial shopfloor prerogatives. The famous sit-down strike at the GM plant in Flint, Michigan, beginning on December 30, 1936, and spilling into the first two months of the new year, set the stage not only for the sit-downs to follow in autos but arguably for their spreading across the larger industrial landscape in 1937. Sidney Fine's careful chronicling of the events of this strike led him to conclude that "it was the speed-up in the view of the principal participants that was the major cause for the GM sit-down strike" (1969: 55). Concerns with speedup and "fair" treatment by foremen sent workers out at Chrysler in March of 1937, where a strong shop steward system led to a better organized strike and a significantly higher participation rate among workers than at GM (Jefferys 1989: 112)

The sit-down tactic spread quickly to other industries. Forty-eight sit-down strikes lasting at least a day took place in 1936; the corresponding number for 1937 was 477. Pencil makers, tobacco workers, rug weavers, pie bakers, and lumbermen, in addition to workers in the more prominent mass-production industries, engaged in sit-downs during 1937 (Fine 1969: 331). Union organizations rode on the coattails of these militant worker actions. At the conclusion of the Firestone sit-down strike, the International president of the autonomous URW asserted that it would "teach the men what an organization can do to settle their grievances." Daniel Nelson's assessment of the strike, however, is probably more accurate: "In fact," writes Nelson, "the sit-down taught the men what they could do for themselves" (Nelson 1982b: 209).

Sit-down strikes were dramatic episodes in the rise of workers' empowered shopfloor voice during this period, but focussing exclusively on them obscures the slow building process—interspersed, to be sure, by occasional breakthrough experiences such as the sit-downs—which led to shopfloor empowerment for workers. This is true for the rise of shopfloor empowerment in autos and rubber, but even more so for such industries as electrical manufacturing, where instances of dramatic breakthroughs in worker consciousness were less common and the process of attaining shopfloor empowerment was a slower, more continuous development away from company-union controlled voice to an independent worker voice.

In electrical manufacturing, more promine.it firms such as General Electric and Westinghouse aggressively challenged outside unions by taking seriously workers' major grievances as voiced through company unions.[15] However, dispute settlement through company unions increasingly left workers dissatisfied, and local labor leaders attempted to build on this dissatisfaction. Militant shopfloor actions occasionally accompanied this process, as in the summer of 1936, when local union leaders led a sit-down strike at the Schenectady GE plant to support the cause of a group of wire-insulation machine operators who were being moved to a measured day work system of payment from a piece-rate system and whose complaints through company-union structures had fallen on deaf ears.[16] But, in general, the process was a gradual one. At times infiltrating the company-union organization to make demands, at other times acting militantly outside its confines, local union leaders slowly built grass-roots support for an independent union voice among workers.

If in struggling to form unions, workers were looking to eventual union recognition and collective-bargaining agreements as ways of achieving shopfloor empowerment, empowerment came prematurely and rested on less formal institutions. Indeed, in many instances an empowered shopfloor presence appears to have been a necessary prerequisite for the attainment of union organization rather than the reverse. Friedlander's (1975) fascinating study of the emergence of a UAW local is revealing in this regard. A local union leader interviewed for the study reported that perhaps 50 percent of the union members joined immediately following certain displays of organized worker shopfloor power (Friedlander 1975: 42). The case study also reveals that in winning recognition and achieving certain contract demands, orchestrated displays of worker solidarity and organized slowdowns were important ingredients.

It was the *activity* of union organizing that produced the breakthroughs in workers' understanding of the latent power of informal work groups, and thereby their shopfloor empowerment. The failure to achieve union recognition and a collective-bargaining agreement within a reasonable time frame probably doomed the long-run viability of workers' shopfloor empowerment. But, there exists persuasive evidence to suggest that union recognition and contract language were neither necessary nor sufficient for the attainment of an empowered shopfloor presence for workers during this period.

That these were unnecessary is nicely revealed by the instances in which shopfloor power was attained in their absence. We often forget just how few workers had attained signed collective-bargaining agreements with employers by the decade's end. In electrical manufacturing, GE succumbed to unionization rather quickly, but Westinghouse did not. Nonetheless, Schatz (1983: 76) notes that Westinghouse was forced to bargain informally with shopfloor collectives of workers if it wanted production to proceed without major disruption. In meat packing, Armour refused to recognize the results of an NLRB representation election in the late 1930s, but workers were still able to get the company's attention and to win certain shopfloor improvements through "stop and go" strikes and a "whistle bargaining" strategy of greeting every grievance with a signal to stop work (Cohen 1990: 305). The activities of informal work groups and solidaristic behavior between such groups empowered workers even in the absence of union recognition and contract protections.

That recognition and contract language were insufficient to ensure the existence of workers' shopfloor empowerment is suggested by a critical look at what precisely was achieved by these features of a mature collective-bargaining arrangement. The attainment of union recognition and a collective-bargaining agreement did not significantly alter the structural features of worker shopfloor representation from those found in company-union settings of the mid-1930s. The formal institutional arrangements of shopfloor representation in the new CIO unions included grievance procedures for dispute resolution and the election of shop stewards and committeemen. Grievance procedures had been a central component in worker representation through company unions since the 1920s. And while compared to shopfloor representation in company unions, union shop steward structures were typically more dense with representatives, the basic structure was certainly nothing new to American industry. Indeed, the structural similarities between the shop steward system of shopfloor representation and that of company unions is striking.

A 1937 UAW pamphlet advising shop stewards on the proper methods for controlling speedups, for example, suggests that this be accomplished "by seeing that the foreman doesn't pit one worker against another, by working with time study in timing the job, and by taking up grievances of speed-up with the foreman" (quoted in Meyer 1992: 107). But, it was not at all uncommon for the empowered company unions of the 1930s to appeal to time study to resolve complaints concerning the pace of production, and company-union structures always provided for requests to foremen concerning worker grievances. The crucial difference in substantive shopfloor outcomes between shopfloor representation through the company unions of the 1930s and the new independent unions was the psychology of increased shopfloor empowerment by workers, and their willingness and collective ability to act on such feelings to win shopfloor demands.

If workers could not rely on structural changes in the institutional arrangements of shopfloor representation for empowerment during this period, neither could they take refuge in contractual guarantees of shopfloor rights. The collective-bargaining agreements of the late 1930s, for those few workers who had attained one, were rather thin documents, covering few conditions of employment other than wages, hours, and procedures for dispute resolution. The one major exception to this is occasional contractual guarantees that promotion, layoff, and transfer decisions would be based on the principle of worker seniority with the firm, thereby removing the autonomy of foremen in decisions concerning the allocation of labor in the plant.

In the absence of agreed-upon contract language governing shopfloor conditions, union grievance procedures per se carried little additional value over similar procedures in company-union settings. Employers, after all, were not legally obligated to accede to workers' demands. Indeed, an analysis of grievances reaching the later stages of the grievance process at one industrial concern following union recognition suggests that management routinely rejected the grievances of shopfloor workers such as requests for electric fans to reduce the heat, gloves to protect hands from dangerous chemicals, rubber boots and hats for workers working outdoors, and the transfer of driving foremen. When, after the introduction of a new technology, workers complained that their new jobs required them to work harder for less money, management rejected their requests for increased pay to compensate for the increased effort (Meyer 1992: 111–17).

The improvements in shopfloor conditions won by workers during these early stages of unionization were won largely on the shopfloor, in day-to-day confrontations, and through informal bargaining over customary practice with shopfloor management. In rubber and autos, where the sit-down was made famous, workers continued to employ this tactic and other shopfloor demonstrations of collective strength to win improvements in such things as the pace of work and the behavior of foremen even after employers recognized unions.

In rubber, after the wave of sit-downs in early 1936, workers sat down on numerous occasions in the coming year to address grievances over production standards, piece rates, and employers' use of nonunion "rate busters" to boost the pace of production. At Goodyear, for example, where labor–management tension was the greatest, tire room workers quit work on the night of March 19–20, corralling foremen and nonunion workers into a makeshift "bull-pen" and holding them there until noon the next day, to protest the appointment of a nonunion man to head a pit crew. The appointment was immediately withdrawn. A sit-down to protest a similar grievance occurred at the same plant almost a year later, in early February of 1937, during which workers could be heard shouting to management, "You may manage this plant, but you don't control it" (Nelson 1988: 206–7, 214–15).

Workers in autos struggled in similar fashion to alter the behavior of foremen, eliminate straw bosses from the shopfloor, and reduce the speed of production.[17] Union committeemen tried to press grievances aggressively through formal channels of dispute resolution, but, as one union member later conceded, "every time a dispute came up the fellows would have a tendency to sit down and just stop working" (quoted in Fine 1969: 321). GM complained to UAW leaders that there had been 170 sit-downs in GM plants between March and June of 1937 (Bernstein 1970: 559). Slowdowns even proved successful on the assembly line according to the recollections of one Dodge worker, who stated "sometimes foremen would jerk up the automatic conveyor a couple of notches and speed up the line. We cured them of that practice: we simply let jobs go by half finished" (quoted in Lichtenstein 1989: 163).

Similar events occurred in many other industries. More militant steel locals, such as the one at Inland Steel in Chicago, utilized "a series of strikes, wildcats, shut-downs, slow-downs, anything working people could think" to secure demands, according to one worker (Lynd and Lynd 1973: 107–9). At Armour, 300 workers in the pork-killing department walked off their jobs in January of 1940 to protest the company's plan to increase the number of hogs to be slaughtered per day (Cohen 1990: 307).

The substantive impact on shopfloor conditions of workers' empowerment during this period remains largely unexplored terrain. Copious anecdotal evidence exists. For instance, a worker at the Fisher Body No. 1 plant, the stronghold of the Flint sit-down strike earlier in the year, reported in March, less than a month after the strike's conclusion:

> The inhuman high speed is **no more**. We now have a voice, and have slowed up the speed of the line. And [we] are now treated as human beings, and not as part of the machinery. The high pressure is taken off . . . .
>
> (quoted in Fine 1969: 328)

There is also the occasional piece of hard evidence, such as Clark Kerr's estimate that labor productivity at GM fell by roughly 10 percent following unionization in the late 1930s (cited in Herding 1972: 29).

However, even better evidence can be marshaled. There exist data making it possible to relate changes in both industry injury rates and labor productivity over the late 1930s to proxies for the extent of worker shopfloor empowerment across industries at the end of the period.[18] The distinguishing feature of worker voice during these years compared to its counterpart in the 1920s is that changes in shopfloor conditions were not limited in the 1930s to improvements that would prove mutually beneficial for workers and firms. The distribution of shopfloor power was altered in workers' favor during this period. Injury rates should have improved in those industries where workers' shopfloor organization was strongest, and, in contrast to the earlier period, productivity growth should have been lower, as empowered workers began successfully to address the burning issue of the pace of production.

Our analysis of the history of rising worker empowerment during these years suggests that an empowered shopfloor presence was attained by workers in the process of struggling to form independent unions. We therefore proxy shopfloor empowerment by union concentration—the extent of contract coverage among an industry's labor force. However, union contract coverage across industries during the late 1930s is an imperfect proxy for the power of workers' shopfloor voice because, as we argued above, possession of a collective-bargaining agreement was neither necessary nor sufficient for the possession of an empowered shopfloor presence by workers during this period. Instead, we utilize a measure of union concentration for the year 1941, which allows us to capture many of those organizing drives of the 1930s that resulted in immediate worker shopfloor empowerment but which only later yielded employer recognition and a signed contract.[19]

There nonetheless remain significant limitations with the measure we have chosen. The measure fails to capture those instances in which workers attained an empowered shopfloor voice but no collective-bargaining agreement as of 1941. It also fails to capture those situations in which a union was organized, recognized by the employer, and a signed agreement was entered into, but the power of workers' shopfloor voice was not fundamentally altered in the process.

The extent of industry strike activity might be thought of as an alternative proxy for worker militancy and empowerment, but strike activity is a more problematic measure of empowerment than union concentration. Strike activity may reflect employer recalcitrance to unionization, and therefore represent a defensive, as opposed to militantly offensive, posture on the part of workers. Moreover, because BLS strike data only account for strikes that have a duration of at least one day, recorded strike activity fails to capture the "quickie" strikes that were so important for building and maintaining shopfloor empowerment during this period. Finally, strike activity may have an effect on injury rates and labor productivity quite apart from its association with worker empowerment, thereby complicating the interpretation of results.

Our hypothesis, then, is that changes in working conditions were comparatively more favorable for workers in those manufacturing industries where union concentration was the strongest in 1941, the latter standing as a proxy for the extent of worker shopfloor empowerment in the late 1930s. The first row of Table 3.1 reports on the analysis of injury rate changes across industries. It presents the results of a simple correlation between changes in industry injury rates over the late 1930s and union concentration. Injury rates declined between 1936 and 1940 by an average of 12 percent for our sample of industries. Although the larger the union concentration measure, the larger is the decline in injury rates, the association is not statistically significant. This result is unchanged with the introduction of a variable capturing the intensity of strike activity across industries.[20]

The introduction of other control variables might well alter these findings. The rate of mechanization is one such variable that immediately comes

Table 3.1 Suggestive evidence on the relationship between independent unions and trends in productivity and injury rates (sign and significance reported)

Regression analysis

	Dependent variable	Independent variables		N
		Union$_{1941}$	Injuries 1937–9	
(1)	Injuries$_{1936-40}$	–		24
(2)	Productivity$_{1937-39}$	–***		13
(3)	Productivity$_{1937-39}$	–***	–	13

* significant at the 0.10 level (one-tailed test)
** significant at the 0.05 level (one-tailed test)
*** significant at the 0.01 level (one-tailed test)

to mind since it is likely to be correlated with both union concentration and injury rates. Union wage increases may encourage firms to further mechanize production in an effort to shed labor. And although mechanization's impact on the injury rate may be either positive or negative (as discussed in chapter 1), it is unlikely to be neutral. Unfortunately, however, data on horsepower per worker which would allow us to control for changing mechanization across industries were not available for this time period.

The last two rows of Table 3.1 contain the results of an analysis of changes in industry labor productivity during the period. The labor input measure was adjusted for worker-days lost due to strikes, and so the productivity measure is output per actual worker-days worked.[21] Labor productivity increased between 1937 and 1939 by an average of 6 percent for our sample of industries. The results presented in the second row reveal that union concentration in 1941 is very significantly associated with reduced labor productivity growth in the late 1930s. When various strike intensity measures were added as control variables to the estimated relationship, the union effect remained robust, while the strike intensity variables were always insignificant.[22]

In the third row we control for changes in industry injury rates over the period, and find that the effect of union concentration on productivity growth operates independently of the impact of the injury rate on productivity. Compared to the results for the early 1920s, and consistent with the results for the late 1920s (reported in chapter 1), improvements in injury rates no longer appear to yield a significant enhancement in productivity growth.

The results of the productivity analysis are consistent with our hypothesis that worker empowerment during the 1930s resulted in a significantly decreased work pace, and, consequently, a reduction in the rate of productivity growth. Of course, we cannot rule out the possibility that the observed relationship between union concentration and productivity growth suffers from spurious correlation. We note, however, that a number of competing explanations for these findings can be confidently ruled out by virtue of our having chosen a reasonably short period of time for analysis, during which the most important change in shopfloor governance was arguably the rise of empowered worker voice.

Moreover, some competing explanations can be subjected to empirical testing. For example, unions are more successful at organizing industries where average establishment size is large, and yet larger-sized establishments may also possess superior resources with which to produce productivity improvements. Failing to account for the average size of establishments in the industry may produce a left-out variable bias in the estimated union effect. However, when a measure of average establishment size was added to the estimating equations there was no appreciable change in the significance of the estimated union-concentration coefficient.

While admittedly suggestive, we conclude that the results of this analysis are supportive of the claim that an empowered worker voice accomplished significant improvements in shopfloor conditions for workers during the late 1930s.

Workers' ability to parlay their shopfloor empowerment into improved shopfloor conditions faced a number of significant challenges from a host of powerful forces as the decade of the 1930s drew to a close. The Supreme Court, in its Fansteel decision of 1939, held that dismissals of sit-down strikers did not constitute an unfair labor practice, thereby sanctioning such dismissals by employers. And the CIO leadership began actively to discourage shopfloor actions such as sit-downs and slowdowns, especially once contracts were signed with employers.

In January of 1939, following a series of militant work stoppages in meat packing, the PWOC leadership issued a warning to local unions to attain the authorization of the national office before engaging in any form of strike activity; never to permit work stoppages in plants that possessed a contract with the employer; and to use the grievance machinery to settle disputes. (Ironically, one of the unauthorized strikes that led to this warning occurred at the Sioux City plant of Cudahy in late December of 1938, where stewards had orchestrated a fifty-minute work stoppage to protest the slow pace of the grievance procedure (Brody 1964: 179).) Similar warnings were issued by the labor leadership in steel, autos, and elsewhere.

In fact, though, few international unions were in a position to enforce fully their demand that locals forgo militant displays of shopfloor activity to win shopfloor improvements. In autos, for example, after much pressure from GM to eliminate workers' autonomous shopfloor expressions, the UAW agreed to contract language in 1937 stipulating that no work stoppages were to occur before use of the grievance procedure had been exhausted, and even then only when international officers had given their approval. However, as Fine notes, the agreement "implied a degree of discipline in the union that did not exist at that time" (1969: 325).

The factionalism that erupted in the leadership of the UAW during this period was in part a disagreement about how to handle expressions of shopfloor militancy by workers. The fighting among top leadership in the union only promoted workers' local power, so that the "discipline" that Fine thought was lacking in 1937 was still quite palpable in the early 1940s, as the UAW president at that time, R. J. Thomas, revealed in saying, "When the men are on strike, what can you do? If you don't authorize it, they'll go out anyway" (quoted in Lichtenstein 1982: 17).

Although the shopfloor quickie strikes and slowdowns of the late 1930s were significant in allowing workers to win improvements in shopfloor conditions, and in enhancing the probability of winning union recognition and a collective-bargaining agreement, they also possessed several drawbacks for workers, especially once a contract with the employer had been won. For a start, the benefits of workers' shopfloor actions typically failed to redound equally to all workers, while the costs, in the form of a lengthy strike or the failure to reach beneficial contract demands, were spread more evenly among the workforce. In addition, because there was often little strategic planning of such events, and little communication among the larger contingent of workers concerning the reasons for and goals of the actions, militant shopfloor displays were sometimes heady responses to mere rumors, and sometimes simply purposeless. Nelson notes, for example, that a common complaint of management in the rubber industry during such displays "was that it was impossible to negotiate because no grievances had been presented" by the workers (Nelson 1982b: 213). Finally, militant shopfloor actions threatened the stability of the new union organizations. Companies looked to unions to rein in shopfloor militancy, and began making success in this area a prerequisite to employer recognition of and willingness to bargain with workers' union representatives.

This tension between worker shopfloor militancy and the organizational stability of unions would be at least partly resolved by the events of the 1940s. Signs of this resolution were visible even by the close of the decade of the 1930s. As a consultant to Allis-Chalmers in 1940, Don Lescohier, a former John R. Commons associate, reported that workers posed a serious challenge to managerial authority on the shopfloor of this company. The workers' strategic use of the grievance procedure came in for much criticism in the report. Most worrisome to its author was:

the character of the grievances it [the union] has been bringing in, its attacks upon the disciplinary controls of the foremen and superintendents, and its repeated defiance of company authority in the shops. . . . Union violations of the contract have consisted principally of invasions of management's control of the shop and grievance tactics designed to stretch the meaning and widen the interpretation of the Agreement.

(quoted in Meyer 1992: 118)

This would be the character of workers' shopfloor actions and the basis of workers' empowered shopfloor voice for almost two decades to follow.

CONCLUSION

By the end of the 1930s semi-skilled workers in many of America's mass-production industries had come to possess a degree of collective power over shopfloor conditions that would have been unimaginable during the relatively powerless years of the company-union movement. The fundamental change in shopfloor governance during this period was not so much in the formal institutional structure of shopfloor representation as in the consciousness of workers, and, in particular, in their discovery of the latent power that lay in the cooperation of informal work groups and militant shopfloor tactics.

The rise of this empowered shopfloor voice was historically contingent on the depression. The depression exposed the limitations of the company unions of the 1920s by provoking a deterioration in those shopfloor conditions not adequately regulated by this form of worker voice. The depression also provided a context in which a collective worker voice in industry was complementary to the larger goals of the nation, thereby encouraging government support for unions.

Company unions contributed to the rise of an empowered shopfloor voice for workers as well, in that they gave workers and worker representatives practice in the fundamentals of collective worker representation and negotiation with management. The attainment of shopfloor empowerment, however, was largely the work of workers. Through the activity of union organizing, workers built informal shopfloor organizations that vied with shopfloor management for control over work pace and other features of shopfloor production.

4

LABOR–MANAGEMENT DISPUTES IN MEAT PACKING, 1936–41

Workers in the packinghouses had become restive by the late 1930s. They had suffered through the speedups and capricious behavior of foremen of the depression years, and had been enticed into both company unions and the decades' old Amalgamated Meat Cutters and Butcher Workmen (BW) of the AFL following the NIRA. In the fall of 1937, Van Bittner, the Steel Workers Organizing Committee (SWOC) midwest director, prevailed on John L. Lewis to let him take on the task of organizing the packinghouse industry. The Packinghouse Workers Organizing Committee (PWOC) was formed, modeled after the SWOC, and the industrial organizing drives of the CIO came to the meat-packing industry.

This chapter utilizes the dispute files of the Federal Mediation and Conciliation Service (FMCS) to explore the shopfloor discontents, union-organizing activity, and shopfloor empowerment of workers in meat packing during the late 1930s.[1] The dispute files of the FMCS are attractive for this purpose because they offer insights into labor–management conflict during the period that are unavailable from sources such as published strike data. Less than one-third of the disputes analyzed in this case study appear in recorded Bureau of Labor Statistics strike statistics. Moreover, in many cases the FMCS dispute files contain details of the events leading up to a dispute, the specific nature of the dispute, as well as its formal resolution. All labor–management disputes on file involving plants of the "big four" packinghouses—Wilson, Swift, Cudahy, and Armour—between 1936 and 1941, inclusively, were reviewed.

Figure 4.1 presents a breakdown, based on several criteria, of the seventy-four recorded FMCS disputes at plants of the "big four" packinghouses during this period. Looking first at the causes of the disputes, we see that roughly 10 percent were the result of workers' discontent with shopfloor conditions. Six out of the seven shopfloor disputes involved work pace or work load issues, and in three of these instances a recent speedup was cited as the specific causal factor. The remaining dispute made reference only to "intolerable working conditions."[2] The files make it clear that these expressions of shopfloor discontent often led to worker demands for union

Total disputes

74

Cause of dispute

Recognition/contract[a]	Wages and hours	Working conditions	Other[b]
61	1	7	5

Type of dispute

Strike	Threatened strike/controversy
19	55

Type of strike

Sit-down	Conventional
5	14

[a]This category includes, in addition to demands for formal recognition and contract, company failure to respect union grievance procedure, to concede to closed shop, to bargain in good faith, and to abide by verbal agreement.
[b]These were predominantly concerned with the failure of fellow workers to pay union dues.

Figure 4.1 Labor–management disputes in major packinghouses, 1936–41

recognition, a written contract, or a grievance procedure with a quicker, more equitable resolution to workers' shopfloor concerns. It is not, therefore, unreasonable to presume that some of the disputes involving worker demands for union recognition or a signed collective-bargaining agreement may also have had shopfloor discontent as a motivating factor.

Interestingly, only one of the recorded disputes during this period was precipitated by worker demands for better wages or hours. A portion of the disputes involving union recognition or a written contract may have had wage and hour demands as a motivating factor, but in only one instance were these issues cited as the immediate cause of a labor–management dispute. This reveals the importance of nonmonetary issues in workers' concerns during the period.

Figure 4.1 also contains a breakdown of disputes by type—specifically, whether the dispute involved a strike, and if so, whether or not it was a sit-down strike. Less than a third of the recorded disputes took the form of a strike, suggesting that workers at the "big four" meat-packing plants were not very strike prone during these years. Further supportive evidence for this conclusion emerges from BLS data on strike intensity (i.e., the percentage of the industry labor force involved in strikes during a given period). The meat-packing industry was well below the average strike intensity for the sample of industries used in our empirical work in chapter 3, for example. Interestingly, however, the files also reveal that over a quarter of the strikes involved workers sitting down on the job, suggesting that a segment of packinghouse workers could indeed behave militantly when provoked.

Is there any correlation between militancy and shopfloor discontent? Disputes which involved shopfloor discontent as the primary causal factor resulted in strikes in all but one of the seven cases. These strikes tended to be more spontaneous in nature; they often began without the authorization of the union leadership, and, in one instance, resulted in workers sitting down on the job.[3] The sit-down strike took place at the Armour plant in Kansas City, Kansas, in 1938, and began as a protest over a speedup.[4] It is atypical of the disputes involving shopfloor conditions only in that it resulted in workers occupying the plant. In most other respects this dispute is illustrative of the militancy of workers' actions in response to shopfloor discontent.

The Armour packing plant was the largest of the packinghouses in Kansas City during the late 1930s, with a workforce of over 2,000. The Packing House Workers Industrial Union No. 232 had established a presence in the plant—playing a prominent role, for example in grievance resolution—over a year before the sit-down strike of September 1938. The union did not possess a written agreement with the company. In late August of that year, plant management announced an increase in production standards—the "kill" rate—which would take effect from the beginning of September. After a week of struggling under the new production standards, six workers in the hide cellar requested extra help, complaining that the work was too much for them. When the company refused to comply with the workers' request, the men sat down. The workers were told to collect their paychecks and leave the plant, and were promptly replaced by six new workers. Upon discovering that the six new workers took 50 percent longer to do the work of the previous six, the company promptly put eight men on the job the following day.

A hearing was held on September 8, the day after the sit-down incident, before a grievance committee which heard testimony for roughly six hours from the disgruntled workers. The committee decided that the men should be returned to their jobs and, in addition, paid for the time they spent testifying before the committee, a common practice at the plant. The company conceded to the first request, but refused to pay for the workers' time before the committee. When, on the following day, the workers' paychecks did not reflect their time giving testimony, a group of workers, including those from the grievance committee, marched through the plant calling for a general sit-down strike. Seventeen hundred workers stopped work and a portion occupied the plant on Friday, September 9, 1938, in the first general sit-down strike in Kansas City's history.

It became clear by the end of the day on Friday that the leadership of the PWOC had approved of the strikers' actions. The Friday *Kansas City Star* reported that Don Harris, chairman of the PWOC in Chicago, was contemplating a sympathy strike in some of the sixteen Armour packing plants across the country, and quoted him as saying, "We believe a grievance of one worker is a grievance of the entire union" (*Kansas City Star* 1938a: 1). The union made immediate plans for feeding the sit-down strikers by setting up a kitchen at CIO headquarters across the street from the plant. Inside the plant the mood among workers seemed lighthearted as card games were in evidence and "Negro" spirituals could be heard. Orville Ussery, president of the local CIO packinghouse workers' affiliate, counseled the workers:

> Now, fellows, it looks like we'll be in here for quite a while yet. . . . We'll have to maintain strict order. Committees will be appointed to handle the food and other necessary matters, but we must maintain discipline. Nobody will be allowed outside the plant. . . . And no beer, please.
>
> (*Kansas City Star* 1938b: 1)

On Saturday, as the workers settled in, the company expressed concern about slaughtered meat that had not been moved to refrigerator storage. It reported that $27,600 worth of meat had spoiled on Friday and that perhaps the same value of meat was likely to spoil if not moved to coolers immediately. (Payment of lost wages to the original sit-downers during their testimony to the grievance committee amounted to $22.09 in total.) The striking workers refused to do the job, but the local union leadership allowed management access to the meat for the purpose of moving it into storage. Foremen moved the salvageable meat into coolers and sent the remainder to a local soap company, where it was made into soap, grease, and oil. During the moving process, a dozen or so workers taunted foremen about the weak effort they seemed to be putting forth, which provoked the company to demand a two- to four-week suspension for these workers as punishment.

Eight hundred of the sit-downers remained in the plant as of the following Monday, and Don Harris reported to the press that he had received telegrams of support from local unions at other Armour plants, and that they were willing to call out their workers in a sympathy strike upon the request of the PWOC chairman. The workers' resolve, the leadership's threat, or perhaps the mayor's efforts to resolve the strike convinced the company to agree to binding arbitration of the dispute. It was agreed that the arbitration committee would consist of one company representative, one union representative, and one neutral party to be mutually agreed upon by company and union. In the event that the company and union could not agree on a neutral representative, the mayor would appoint one.

The agreement to resolve the dispute in this way was drawn up on Monday evening, September 12, and the sit-down strikers left the plant and returned to their jobs the following day. As the agreement was being drawn up, the plant manager, P. A. Dett, was heard saying to Ussery, "You boys want to be sure and pitch in and clean up the place." Ussery was reported to have smiled and replied, "The boys would get the place going again all right"—a not so subtle indication of the sense of empowerment workers felt as a result of their victory and the likely consequences of this empowerment for labor–management shopfloor relations in the days and weeks to come (*Kansas City Star* 1938c: 1).

One of the interesting features of the meat-packing industry during this period is that the old AFL union—the Butcher Workmen—competed with the PWOC for union members. The FMCS dispute files identify the union involved in each dispute, thereby allowing a comparison of the activities of the two contending organizations. Figure 4.2 presents a breakdown of the disputes by union affiliation using the criteria of Figure 4.1. The breakdown reveals that all of the working conditions disputes involved the PWOC. By the available measures of militancy, workers associated with the PWOC were also far more militant than those organizing under the banner of the BW. Less than 20 percent of the BW disputes took the form of strikes, whereas more than 30 percent of the PWOC disputes resulted in strikes. Moreover, all of the sit-down strikes involved the PWOC. Roughly 40 percent of the PWOC strikes took the form of sit-downs, and sit-down strikes composed more than 10 percent of the total PWOC disputes.

In contrast to the BW, the PWOC validated workers' shopfloor discontents and facilitated collective demands for a resolution to workers' most

United Packing House Workers

Total disputes

40

Cause of dispute

Recognition/contract	Wages and hours	Working conditions	Other
29	1	7	3

Type of dispute

Strike	Threatened strike/controversy
13	27

Type of strike

Sit-down	Conventional
5	8

Butcher Workmen

Total disputes

34

Cause of dispute

Recognition/contract	Wages and hours	Working conditions	Other
32	0	0	2

Type of dispute

Strike	Threatened strike/controversy
6	28

Type of strike

Sit-down	Conventional
0	6

Figure 4.2 Union affiliation and labor–management disputes in major packinghouses, 1936–41

pressing shopfloor concerns. This is nicely illustrated in the FMCS files by the details of several instances in which a plant was originally organized by the BW only to be later represented by the PWOC. Events at the Armour plant in Fargo, North Dakota, during 1938 and 1939 are revealing. In May of 1938, Patrick E. Gorman, President of the BW, contacted the Director of Conciliation, John R. Steelman, to complain about the failure of management at the plant to recognize the BW local there.[5] Gorman claimed, in particular, that management had failed to post a verbal agreement between the company and the union, behavior which, he warned, could provoke the union to strike. (The posting of verbal agreements was the precursor to more formal written agreements at many companies during this period.) Steelman dispatched a conciliator to the plant, and the conciliator convinced the company to post an agreement by noting that the company regularly posted similar agreements between itself and the former company union.

On June 15, the union presented the company with a detailed agreement which contained language governing pay rates for specific job classifications in the plant, work hours, break periods, paid vacations, and a no-strike clause committing the company and union to binding arbitration of disputes. The language covering the grievance procedure was extremely vague, however, stipulating only that the Business Agent of the local must be given a chance to adjust disputes before they are sent to arbitration. There was no mention of shop stewards and no reference to an internal system of dispute resolution.

Later, in August of the same year, Albert Winters, the local president, complained in a letter to the Director of Conciliation that the company was refusing to deal with the workers as a collective. In particular, the complaint was that workers had been instructed by the company to take up grievances individually with their respective foremen, presumably without union representation. (The Business Agent of the local was apparently the only union representative allowed to present workers' grievances, and he was required to do so to the Superintendent.) Steelman again dispatched a conciliator to the plant. The conciliator reported that the plant manager was an obstructionist, and decided that the best strategy would be to contact Gorman and request that someone be sent down from the national office of the union with the capacity to deal with a plant manager of such character. This is how things were left.

In April of 1939, the NLRB certified a local of the United Packing House Workers Industrial Union as the duly elected bargaining agent for workers at the Armour plant in Fargo. By August of that year, another dispute at the plant was brought to the attention of the Director of Conciliation.[6] This time, neither the local union in Fargo nor the national union leadership initiated contact with the US Conciliation Service. The PWOC was less apt to turn to the government for aid in resolving disputes, preferring instead to allow workers to handle matters locally. When the conciliator arrived, he found a strike in progress. The union claimed that the company had unfairly discharged two workers; the company claimed that the dismissals were fair because the workers had failed to carry out a foreman's orders and were generally insubordinate.

The specific details are not in dispute. Clarence Wheeler, chief steward of the local, and a fellow co-worker, refused to follow a foreman's orders and protested against "the inhuman and vicious speed up" (*North Dakota Union Farmer* 1939: 1). When the company dismissed the men and refused to address the dispute directly, or agree to send it to arbitration, the strike was called. Twenty-five men were charged with rioting on the first day of the strike, August 21. On September 1, the company offered to reinstate all strikers with the exception of the two men who had disobeyed orders, and warned that the striking workers had better return to work or their jobs would be in jeopardy. The strikers refused.

On September 18, Don Harris, Director of the PWOC, sent a letter to the secretary of the local in which he renewed his support for the strike, promised continued financial assistance to striking workers, vowed to stop shipments from the Fargo plant to other Armour plants, and, although unwilling to pull out workers at the other Armour plants in a sympathy strike, suggested that the union wait for the national scene "to develop whereby we would force a complete victory from the Fargo management."[7] The strike lasted seventy-nine days in total, from August 21 to November 7. The resolution involved the loss of seniority for those workers who remained on strike and formal discharge of the two men who had disobeyed the foreman's orders.

On most issues of importance to the empowerment of workers in meat packing during this period, PWOC members and their union leaders were militant in their demands. Compare the response (discussed above) of the BW local at the Armour plant in Fargo when in the fall of 1938 the company refused to recognize a union shopfloor grievance structure, with the response of the PWOC local at the Sioux City, Iowa, packing house of Swift & Company when, that same fall, the plant management behaved in similar fashion.[8]

On September 29, 1938, management at the Swift plant refused to take up a grievance being presented by a grievance committee of the union. This was consistent with the company's long-held policy of having workers express their grievances to foremen individually. The workers chose to strike in response. The strike began with 300 workers sitting down and refusing to work upon hearing of the company's refusal to meet with the union grievance committee. After eight hours, 500 workers, including the original sit-downers, left the plant to pursue a conventional strike against the company. It would last 119 days and attract national attention.

During the course of the strike, the Regional Organizer of the PWOC and the local union president, among others, were arrested, a group of sixty-five workers was indicted on charges of conspiracy, rioting, and carrying concealed weapons, and the National Guard was called out to maintain the peace. Local banks refused to honor checks of the PWOC, bail bondsmen refused to grant bonds to workers and union leaders who were arrested, telephone wires were tapped, and stool pigeons attempted to infiltrate the union organization. On January 25, 1939, the company agreed to reinstate all but a core of strikers responsible for the original sit-down.

Did the struggles of this period, many of which preceded union recognition and collective-bargaining agreements, lead to empowerment for workers on the shopfloor? The FMCS files do not provide a sufficient longitudinal sample of disputes at particular work sites which would allow us to track with some confidence the evolution of shopfloor labor–management relations during this period. There is, nonetheless, a perceptible change in the character of disputes over these years, away from the more desperate expressions of discontent in the early period and towards a more deliberate and controlled use of shopfloor tactics such as the slowdown, later on, which is suggestive of the evolution towards a higher stage of collective and strategic empowerment. Events at the Cudahy plant in Kansas City in the summer of 1941 serve as a useful illustration of these later shopfloor tactics.[9]

In early July, Sidney Hillman, then at the Office of Production Management in Washington, received telegrams from both the plant superintendent and the president of the local union indicating some tension in the plant concerning the company's respect for the union. The local union president complained to Hillman about the company's failure to meet with an authorized bargaining committee of the union; the plant manager responded that the union was engaging in "periodic slow-downs," and that local officers were "promoting discord and a lack of discipline between supervisors and employees."[10]

On July 25, the union announced its demand for a union shop clause in the contract, and threatened a strike if the demand was not met. Director of Conciliation Steelman dispatched a conciliator in response to this threat, who then reported back to Steelman on July 29, that a slowdown in production was taking place, presumably as part of the attempt to win the union shop. E. A. Cudahy, Jr. fired off a heated letter to Steelman on July 30, complaining of the organized slowdown stating that

> the gangs are all in this a.m., but resorting to the same slow-down tactics. I don't think it is unreasonable to estimate that these present tactics are costing us at the rate of three to four hundred thousand dollars a year. [The leader of the PWOC] must take some courageous action within his own organization or a very radical element is going to get in entire control of the packing house workers' organizing committee.[11]

The Cudahy dispute suggests that by 1941 workers had attained a degree of shopfloor empowerment sufficient to collectively control such shopfloor conditions as the pace of work, and to do so in order to achieve larger strategic goals such as union contract demands. A similar instance of the strategic use of the slowdown to, in this case, achieve union recognition can be found in a dispute at Wilson during the same year,[12] lending further evidence to the claim that shopfloor empowerment was as at least as important to the establishment of independent unions as formal union structures were to shopfloor empowerment.

CONCLUSION

The FMCS files on meat-packing disputes during the late 1930s offer evidence in support of a number of claims made in the last chapter. Shopfloor discontent was a prominent cause of labor–management conflict during the period of the industrial organizing drives of the 1930s. In the case of meat packing, workers' shopfloor concerns rivaled those of wages and hours among disputes where a specific causal factor other than union recognition was mentioned. Shopfloor discontent also seems to have engendered a militant response from workers. Expressions of shopfloor discontent in meat packing were more likely to be spontaneous and more likely to involve a strike than disputes involving other issues.

The files also suggest that the PWOC was more committed to addressing workers' shopfloor grievances, and allowing more militant shopfloor actions as both an expression of discontent and a method for controlling shopfloor conditions, than was the BW union. All of the disputes involving shopfloor conditions and sit-down strikes in meat packing during the late 1930s took place in the presence of an active local of the PWOC.

On a more speculative level, the files offer some evidence for the claim that workers became aware of the latent power of cooperative informal work group activity through these militant expressions of shopfloor discontent, and built an empowered shopfloor voice which allowed for the collective regulation and improvement of certain shopfloor conditions by the close of the 1930s. Workers in meat packing appear to have possessed greater collective control over aspects of shopfloor production, such as pace, by the decade's end, and used this control to achieve union recognition and contract demands that went well beyond the shopfloor.

5

INSTITUTIONALIZATION AND DECLINE IN WORKERS' SHOPFLOOR POWER

Industrial relations after World War II in manufacturing are generally viewed as containing a brief period of initial turbulence followed by an extended stretch, beginning in the late 1940s, of relative stability, during which time a "web of rules and regulations" and a set of "mutual understandings" are seen as guiding the actions of employers and organized workers towards the maintenance of industrial peace through the receipt of mutual prosperity. To the extent shopfloor governance is considered in conventional accounts of the period, it is treated in one of two ways: (1) shopfloor conditions are viewed as subject to joint labor–management regulation through the "web of rules and regulations" encompassed in collective-bargaining agreements and grievance procedures (e.g., Brody 1992); or (2) shopfloor conditions are seen as being the sole terrain of management under a "mutual understanding" or "accord" in which workers abdicate control of the shopfloor in exchange for wage increases tied to productivity growth, employment security, and improvements in fringe benefits (e.g., Bowles, Gordon, and Weisskopf 1983).

The appearance of shopfloor governance offered by these accounts does not square with the reality of shopfloor practice. The postwar industrial relations system may have offered a stable set of institutional arrangements for determining wages, hours, and fringe benefits, but it never encompassed a similar set of arrangements for determining shopfloor conditions. A closer look at postwar shopfloor practice reveals that collective-bargaining agreements and grievance procedures were initially underutilized and ultimately inadequate mechanisms for shopfloor governance and that any mutual understanding concerning management's prerogative in production was struck between employers, the government, and the labor leadership, without the consent of rank-and-file workers. Within this rather loose set of institutional arrangements there was a struggle for control over shopfloor custom and practice by labor and management.

The postwar period witnessed first the institutionalization, and then the decline, of workers' shopfloor power in the organized manufacturing sector. The informal system of shopfloor governance that emerged in the early years, in which shopfloor labor and management hammered out custom and practice in a manner that granted significant power to workers over shopfloor conditions, gave way to a more formal contract-and-grieve approach, in which workers' rights with respect to shopfloor conditions were increasingly limited to those specifically spelled out in collective-bargaining agreements. The institutionalization of workers' shopfloor power begins during World War II.

WORLD WAR II AND WORKERS' SHOPFLOOR POWER

The shopfloor power that workers had attained during the industrial organizing drives of the late 1930s had not, by the early 1940s, evolved much beyond the ability to disrupt production. This power manifested itself as a reactive, protest-oriented display of worker discontent with management decisions. Workers needed to transform this form of shopfloor empowerment into an active and constructive force that allowed for the joint regulation of shopfloor conditions with management. It had to blend in with the emerging institutional arrangements of shopfloor governance, to become part of day-to-day shopfloor practice; in short, workers' shopfloor power had to become institutionalized.

Several features of the war period served to move things in this direction. The unions' endorsement of the no-strike pledge during the war placed constraints on workers' expressing shopfloor power through the militant disruption of production. Although not entirely successful, the pledge nonetheless sent a clear signal regarding the low tolerance of union leaders, government officials, and employers for disruptive tactics by workers in production. It was also during the war period that the formal institutional arrangements of what would come to compose the postwar system of industrial relations began to fall into place. Workers experimented with contract language and formal grievance procedures as mechanisms for the joint regulation of shopfloor conditions, and found them to be woefully inadequate. Workers would have to find other outlets for the expression of shopfloor power.

In December of 1941, at a formal meeting of labor leaders and employers, President Roosevelt requested endorsement of his policy to "prevent the interruption of production by labor disputes during the period of the war" (quoted in Glaberman 1980: 3). Labor and management would be expected to resolve disagreements through orderly procedures. Unions would abide by a no-strike pledge and employers would agree to refrain from lockouts. Quickie strikes and slowdowns would not be tolerated, and it would be the business of union leaders to see to it that workers adhered to both the no-strike pledge and the cessation of disruptive expressions of shopfloor militancy.

The labor movement had consolidated forces during the year preceding Roosevelt's request, bringing holdouts such as Ford, Goodyear, Westinghouse, and Little Steel into the union fold, and expanding its ranks by roughly

1,500,000 workers. Despite these gains, however, union organizations remained in a rather precarious position *vis-à-vis* both employers and their membership, and so there was much concern among labor leaders about the possible consequences of abiding by the President's request. Organized workers who were unable to strike would find it impossible to fend off employers' efforts to rid themselves of union interference if the latter were left free to do so. Moreover, union efforts to rein in rank-and-file shopfloor militancy ran the risk of jeopardizing workers' continued support for unions, and thus the dues payments that sustained the organizations. Endorsing the President's request could amount to organizational suicide without some assurances of greater security.

The government responded to these concerns by making existing union organizations more secure. One component in the new security provisions was forcing recalcitrant employers to recognize unions and to bargain in good faith during contract negotiations. Another was the "maintenance-of-membership" formula, whereby upon winning a contract, unions would be guaranteed union dues from the existing membership (and any new members thereafter) for the life of the contract. Maintenance of membership was applied selectively at the beginning of the war, but during the summer of 1942, the National War Labor Board (NWLB) extended the provisions of this policy automatically to any union whose leaders agreed to enforce the no-strike pledge (Lichtenstein 1982: 79). And finally, to facilitate the collection of dues from the union membership, a dues checkoff clause was often added to labor agreements at the direction of the National Defense Mediation Board or its successor the NWLB, which obligated employers to deduct union dues directly from workers' paychecks.

In exchange for union security, many union leaders encouraged workers to forgo strikes and other disruptive shopfloor tactics as means of influencing shopfloor conditions. Unions that failed to put adequate pressure on the rank and file to adopt more orderly procedures for dispute resolution were threatened by the NWLB with removal of maintenance of membership and the dues checkoff. In 1943, for example, the NWLB denied maintenance of membership to Chrysler locals of the UAW because it felt the local leadership was not vigorous enough in opposing a flurry of recent wildcat strikes at these plants (Lichtenstein 1982: 180).

At the same time that workers were being pressured to suspend their shopfloor militancy in support of the war effort, a clearer vision of the formal institutional arrangements by which labor and management would jointly determine the conditions of employment was falling into place. The vision was neither surprising nor new.[1] Organized labor and management would engage in a system of collective bargaining whereby the bargaining agreement would spell out the jointly agreed upon rules and regulations, and a grievance procedure culminating in binding arbitration would act as the adjudicative mechanism for resolving disputes concerning the mutually-agreed upon terms of the agreement. Both parties would be expected to resolve disagreements through orderly contract negotiations and the grievance procedure.

Either the joint regulation of shopfloor conditions were not uppermost in the minds of those responsible for this vision, or, if they were, the vision's limitations in this regard were strategic concerns. Either way, the limitations were glaring. The regulation of shopfloor conditions is terribly difficult to conduct through contract language, and without the strong support of labor leaders and the willingness of employers to engage in negotiations over such matters, what workers tended to get were vague generalities such as, "the employer is responsible for maintaining healthy and safe working conditions" so typical of contracts during these years.

Workers experimented with formal procedures for grievance resolution and found them to be seriously wanting as well. Without protective contract language spelling out workers' rights, employers were not compelled to respect workers' interests in shopfloor disputes. Moreover, the resolution of disputes was extremely slow. Management had apparently taken removal of the strike weapon as a cue to ignore grievances by letting them drift slowly upstairs to higher and higher stages, eventually culminating in a decision by an impartial umpire or in some cases by the NWLB. Lichtenstein reports that by 1943, the NWLB was receiving 10,000 to 15,000 new cases of unresolved grievances per month, and was incapable of reviewing perhaps one-third that number (Lichtenstein 1982: 120).

Many workers responded to being hamstrung in these areas by relying on tried and true methods, such as the slowdown and ultimately department or plant shutdown. The number of strikes (all of them "wildcats" because of the no-strike pledge) rose rapidly between 1942 and 1944, with 1944 witnessing the largest number of strikes in over three decades (Glaberman 1980: 36). Some of the strikes were directed at the perceived limitations of the new arrangements. In 1942, workers at the Brewster Aeronautical Corporation organized a week-long slowdown campaign to protest the NWLB's delay in adjudicating grievances. In 1943, the Chrysler locals in Detroit shut down their respective plants for four days to protest the slow handling of grievances (Lichtenstein 1982: 130).[2]

However, many other workers responded by exploring new and innovative ways of exerting power on the borders of the emerging confines of acceptable procedure. At the Dodge Main plant in Detroit, for example, shop stewards relied on a recently won contractual guarantee that a union representative could observe the company's time study of jobs to influence the custom and practice by which rates were set. Shop stewards, backed by the implied threat of informal work group action, went head-to-head with foremen and lower-level supervisors to chart out a sizeable terrain of jointly controlled shopfloor practice.

There was thus a visible, but far from complete, movement during the war, away from isolated informal work group expressions of militancy, towards more coordinated tactics involving stewards and local union offi-cials, and devoted to influencing norms of custom and practice in shopfloor governance. These developments set the stage for the institutionalized form of shopfloor power that workers would employ in the decade or so following the war.

THE PERIOD OF FRACTIONAL BARGAINING

Developments during the first few years following the war revealed just how much had yet to be worked out on the issue of managerial prerogative in production, and just how anxious employers were to have this issue resolved (Chamberlain 1948). The balance of power in the plant had shifted dramati-cally in labor's favor in the preceding decade. The testimony of a shop foreman before members of a Senate investigating committee in 1945 is illustrative of the degree to which the power balance had changed. In response to a senator's question regarding the foreman's relation with the shop steward, the foreman (a Mr. Bone) replied:

MR. BONE	If the night shift is short a man, he [the steward] even allows me to put an extra man in there, running, say, three men on two machines.
MR. MEADER	You say he allows you to?
MR. BONE	Well, he says it is all right, and he could say No.
MR. MEADER	It sounded as though you couldn't do it without his permission.
SENATOR FERGUSON	He says he couldn't.
MR. BONE	We work together.
SENATOR FERGUSON	I mean you couldn't do it if he didn't want you to.
MR. BONE	He could order the C.I.O. man that he would take, you know, just ease along.
SENATOR FERGUSON	What could he do if he didn't want him to go?
MR. BONE	Well, that man on his machine, he has got to get his production out.
SENATOR FERGUSON	You wouldn't have any authority to change a man's shift at all?
MR. BONE	Not without consulting him.
SENATOR FERGUSON	Without consent, even.
MR. BONE	Yes.

(quoted in Chamberlain 1948: 106)

An executive in the auto industry described the situation more succinctly when he stated, "If any manager in this industry tells you he is in control of his plant, he is a damn liar" (Slichter *et al.* 1960: 561).

Employers in the auto, electrical, and rubber tire industries made the restoration of management's ability to manage production their top priority in the immediate postwar period. Instead of focussing on reining in workers' ability to influence shopfloor conditions through informal shopfloor activities, employers emphasized the collective-bargaining demands of unions in the area of production decisions, perhaps assuming that once the issue of contractual demands was resolved labor leaders could be expected to successfully quell the informal shopfloor actions of workers (Harris 1982: 67).

The labor leadership's war-time endorsement of the no-strike pledge did not imply their willingness to abdicate authority over production matters to management. Clinton Golden and Harold Ruttenberg's 1942 manifesto, *The Dynamics of Industrial Democracy*, which spelled out a plan for moving towards more democratic decision making in steel production, still resonated with many in organized labor. Labor leaders were also aware that discontent over shopfloor conditions had been one of the central reasons for workers' efforts to organize unions, and that improvements in such conditions, albeit largely through informal shopfloor procedures, was a principal result of the 1930s organizing drives. Only a few years had elapsed, after all, since UAW president-to-be, Walter Reuther, had commented, "One of the purposes for which we organized our union was to slow down the assembly lines to a pace that was in keeping with the way a human being ought to work" (quoted in Chamberlain 1948: 284–5).

President Truman convened a Labor–Management Conference in November of 1945, at which prominent members of the business community and union leaders struggled to hammer out an agreement on the institutional arrangements that would guide industrial relations for the postwar period. There was immediate agreement that collective bargaining would be an important component of the postwar system. However, no agreement could be reached on the issue of management prerogative in production and the proper scope of union demands (Harris 1982: 111–18).

In a joint committee on the subject of managerial prerogatives, management members concluded from listening to labor members of the committee that "the field of collective bargaining will, in all probability, continue to expand into the field of management" and further noted that:

> The only possible end of such a philosophy would be joint management of enterprise. To this the management members naturally cannot agree. Management has functions that must not and cannot be compromised in the public interest. . . . Labor must agree that certain

104

specific functions and responsibilities of management are not subject to collective bargaining.

(Chamberlain 1948: 7)

The members of the management group were even forthcoming about which "specific functions" they felt should be left to management's discretion— among them were the organization of work ("the processes, techniques, methods, and means of manufacture"), the job content ("establishing the duties required in the performance of any given job"), and the "determination of safety, health, and property protection measures" (Chamberlain 1948: 154–5). Furthermore, argued the management group, the grievance procedure should be limited to decisions concerning disciplinary actions and the application of the seniority principle. The labor members of the committee rejected such notions, arguing that to exclude labor from participation in these areas was not only inconsistent with collective bargaining, but would likely result in inefficiency of operation as well.

What conference dialogue could not resolve, head-to-head confrontation during contract negotiations did. A partial resolution to the concerns of employers about the extent of union demands in collective bargaining was reached in the contest between the UAW and GM in the latter half of the 1940s. General Motors came to the 1945–6 bargaining round intent on restoring the prerogatives of management in production. The company demanded, among other things, a reduction in the number and powers of union committeemen; the formal processing of grievances in order to reduce continuous bargaining between shopfloor labor and management; management's right to decide "the methods, processes, and means of manufacturing" and to maintain the "discipline and efficiency of employees"; and assurance from the union that there would be no official opposition to speedups in production (Harris 1982: 139–43). General Motors had thus taken it upon itself in this, the first round of bargaining following the war, to confront the issue of managerial prerogative head on.

The UAW countered with demands for the abolition of bonus systems because of their association with speedups; stricter rules governing the allocation of labor in the plant based on seniority; an end to the power of management to set production standards; and an increase in the numbers and powers of committeemen. In making these demands, the union leadership was striving to replace the more militant shopfloor activities of workers with a more responsible, but nonetheless joint, approach to shopfloor governance, involving contractual guarantees and their enforcement by an army of local union officials. However, the eventual deadlock, and ensuing strike, ended in defeat for the union, thereby setting the stage for an approach to contract negotiations which effectively relegated shopfloor issues to the back burner and placed primary emphasis on wages, hours, and fringe benefits.

The outcome of the GM/UAW negotiations in the late 1940s served as a model for much of the negotiations to follow in major manufacturing. No-strike provisions in contracts, which compelled workers to use the grievance machinery instead of striking to win demands, and "management's rights" clauses, which granted management all rights not formally spelled out in contract language, spread rapidly throughout autos and beyond. In the 1947 contract round at US Steel, for example, management won both a management's rights clause and a broad-sweeping no-strike provision.

What emerged from the developments of the late 1940s was a mutual understanding concerning the bounds of collective-bargaining demands, struck between the major players in the postwar industrial relations system—employers, union leaders, and the government. It required that employers willingly enter into negotiations with labor over wages, hours, fringe benefits, and a specified method—namely, seniority—for allocating workers across certain jobs in the plant. The government would support, through various government-financed agencies, the process of responsible collective bargaining. And unions would refrain from making demands that infringed on management's right to "run the plant."

There was one central problem with the vision of postwar industrial relations implicit in this "accord" or mutual understanding: the rather vague understanding reached by union leaders, employers, and the government over management prerogative was not consented to by rank-and-file workers or many shopfloor union officials, and the union leadership was not in a position to see to it that these shopfloor actors fully adhered to the understanding. While more militant exertions of shopfloor power might be significantly reined in, other attempts by workers to influence shopfloor conditions through custom and practice would be much more difficult to control.

The accord was nonetheless important for the impact it had in the area of contractual regulation; contract language would play a very small role indeed over the next fifteen years or so in the determination of shopfloor conditions. The insignificance of contract language governing shopfloor matters is not generally acknowledged in the literature on postwar industrial relations (e.g., Piore and Sabel 1984; Brody 1992). All too often in this literature both the potential for contractual regulation of shopfloor conditions and the extent to which contracts actually did govern shopfloor conditions are overstated.[3]

Postwar collective-bargaining language is often viewed, for example, as spelling out a set of job classifications for each plant, and a detailed description of the tasks and production standards associated with each job, thereby allowing for the joint regulation of working conditions. In fact, though, job descriptions were rarely a matter of contractual stipulation during the early postwar period. And when they were, they were very loose descriptions, sometimes inexact, and often not up to date. Agreements on production standards were similarly vague, typically making reference only to "fairness" and "equity" in the determination of standards. Moreover, management almost always possessed the formal right to alter job descriptions through altered job content—i.e., production standards for a job or the tasks that compose a job.

The desciption by Slichter, Healy, and Livernash of the situation in steel held more broadly across other manufacturing industries where the union's role in formal job evaluation was much less strong:

> Job descriptions were written by the company and reviewed by the union. It was agreed that the existence of the job description did not mean that job content could not be changed by the company. If job content was found in the future to differ from the job description, the description was to be revised. Job content governed the job description and not the reverse.
>
> (1960: 574–5)

The union guarantee of formal influence in this area was thus significantly limited. Unions typically possessed the contractual right to grieve production standards after they had been determined by company engineering departments, and to request a reevaluation of a job's worth if the list of tasks had changed. However, management considered itself the final arbiter of production standards and job tasks, and arbitrators were generally inclined to agree with management on this issue (Slichter *et al.* 1960: 253).

Fortunately for workers, formal shopfloor rights as spelled out in contract language and company policy bore little relationship to existing shopfloor practice during the early postwar period. Strict adherence to the principles of a contract-and-grieve approach to shopfloor governance would have meant a significant diminution in workers' recently won ability to influence shopfloor conditions, and would have elicited a struggle far worse than anything that was witnessed in the immediate postwar period.

What emerged in actual shopfloor practice during the immediate postwar period was neither management's unobstructed right to "manage production" or its right subject only to certain contractual provisions as negotiated in collective-bargaining agreements. Something quite different existed in shopfloor practice. The "quickie strike" had become less common in day-to-day shopfloor relations, and although the slowdown was still occasionally employed in those industries and departments whose production processes left some control over pace in the hands of workers, these older expressions of shopfloor power had become integrated with a more coordinated effort to influence shopfloor custom and practice within the loose confines of the contract-and-grieve approach to industrial relations. Workers' shopfloor power had become institutionalized within a system of shopfloor governance I shall refer to as "fractional bargaining," after the technique for worker shopfloor empowerment it contained (see Chamberlain 1948).

Fractional bargaining was a decentralized and strategic form of shopfloor governance involving foremen, shop stewards, and informal work groups. It rested fundamentally on two underlying features of production in the immediate postwar period: (1) company policy and collective-bargaining agreements left a sizeable scope for custom and practice to dictate the shopfloor conditions of production; and (2) the determinants of custom and practice were largely the terrain of shopfloor labor and management. Under the system of fractional bargaining, foremen and shop stewards possessed the ability, though not necessarily the right as defined by either company or union, to strike extra-contractual deals governing shopfloor conditions. Fractional bargaining, as a technique for the exertion of workers' shopfloor power, has therefore been aptly described by Hyman as "the unauthorized pursuance of demands backed by unofficial sanctions" (Hyman 1972: 62).

Kuhn's (1961) survey of twenty major manufacturing firms during the mid-1950s revealed that the internal grievance procedure was being used by workers as a mechanism to facilitate shopfloor bargaining over workplace concerns. In his interviews with foremen and shop stewards, Kuhn was informed that roughly 80 percent of all grievances were resolved by the give-and-take efforts of these two groups of actors (Kuhn 1961: 27). Many extra-contractual workplace issues were resolved in this fashion well into the mid-1950s, and even certain contractual rights possessed by management were eliminated through this technique. Strauss writes:

> In the typical company throughout the 1945–55 period there developed a whole series of informal relationships between union and management. Grievances were often handled on a "problem solving" basis without much reference to the specific terms of the contract. Foremen and stewards, superintendents and committeemen, were permitted and even encouraged to reach private unwritten understandings or "bootleg agreements" which in effect modified the contract . . . . On the whole, this was a type of guerrilla warfare in which the union had all the advantages of terrain.
>
> (1962: 86)

Kuhn's description of the distinction between theory and practice in the shopfloor grievance process is also revealing:

> The men who handle grievances for management and the workers are typically negotiators, daily adjusting the provisions of the collective agreement and plant rules and codes to fit local needs and desires of people involved in the production process. . . . [There is] a broader range of activities in the grievance process than the union and company negotiators contemplated when they established the grievance procedure. They designed the grievance arrangements in the shop to secure compliance with the agreement. However, the altered power structure in the workshop does not allow shop management to carry out its programs solely by itself, as originally conceived. The

108

grievance process in reality includes all the varied activities of representatives of union and management in the plant that establish,
maintain, or apply the local shop terms of work. Thus defined, the
grievance process encompasses more than the judicial forms and theory
allow. It is, in fact, continuous shop bargaining, with all the benefits
and faults implied therein.

(Kuhn 1961: 77)

The benefits for workers from fractional bargaining stemmed, in part, from
the fact that stewards were in a better position to utilize such a system than
were foremen. Many stewards spent their entire day in contract administration, and so knew the provisions of the contract—both what it included and
what it did not—better than foremen, who had many other responsibilities.
As one line steward put it, "Any steward who knows his stuff can talk rings
around a foreman. If he says the foreman's wrong and talks enough, whether
he's entirely accurate or not, he's apt to buffalo him" (Kuhn 1961: 29).
A typical foreman's response was:

You're tied up with stuff in the shop all the time and you don't have
time to study the contract. I don't study the grievances and arbitration
cases at home like I guess I could. But then the contract can be very
confusing, and you're not sure whether you'll have the right interpretation or not.

(Kuhn 1961: 29)

Foremen were also vulnerable to workers' strategic use of the grievance
process because industrial relations departments took note of foremen whose
subordinates filed an inordinate number of grievances (Sayles and Strauss
1967). Thus, sometimes just the threat of filing grievances could get
foremen to budge on issues of custom and practice. At other times, workers
would flood the process with grievances over an issue of minor concern and
use the lack of a speedy resolution as an excuse for a "quickie" strike, during
which the more pressing, and typically noncontractual, concern could be
addressed (Seidman *et al.* 1958).
The form and extent of fractional bargaining varied between industries,
with rubber and electrical equipment workers able to exert relatively more
shopfloor power over the period through this technique than their fellow
workers in, say, oil or chemicals. Sometimes there was even a large variation
among firms within an industry. In autos, for example, GM was quick to
adopt a narrow and legalistic approach to unionized workers' rights.
Industrial relations became a top management function; well-defined rules
and regulations guided daily decision making; and the opportunity for
decentralized bargaining was held to a minimum (Harris 1982: 28–9). The
story at Chrysler could not have been more different.
Chrysler possessed an avid anti-union stance well into the early 1950s,

refusing to agree to the dues check off, for example, long after other employers had conceded to this measure of a stable relationship with unions. The effect of this stance was that workers at Chrysler maintained an organizing mentality and a militancy that became institutionalized in the union structure, one example of which was an incredibly strong steward system, in which one of the stewards' jobs was dues collection. Chrysler also possessed an extreme centralization of managerial decision making during the period through the early 1950s, but without the proper bureaucratic form to make it feasible, thereby creating an environment ripe for shopfloor custom and practice struggles and fractional bargaining gains by workers well into the mid-1950s (Jefferys 1986).

Workers at the Dodge Main plant in Detroit, for example, utilized shopfloor bargaining, backed by a willingness to slow down and even shut down production, to maintain their sense of "fair" production standards. Stewards hammered out production standards jointly with foremen during the process of each model change. If stewards were less than fully successful, or needed some added support before a final decision was reached, slowdowns or sit-downs were invoked. Using such fractional bargaining tactics, workers at Chrysler possessed significant control over job content well into the 1950s. In the body shop, one worker reported in 1949:

> we ran the job just as we saw fit and worked 40 or 45 minutes each hour. We'd get production ahead and then sit down to talk or rest or kid around. We never worked more than 45 minutes out of an hour, and sometimes, only 35.
>
> (quoted in Jefferys 1986: 112)

The same was true for assembly-line workers, as one worker recalled:

> the standards [on the assembly line] were pretty much decided by the workers on the job. We would decide how much we could do. . . . We didn't have any relief—and we wanted it that way. We made our own relief. Without jeopardizing my job I could make 15 minutes for myself every hour. The foremen knew what was going on. But there was time to do good quality work.
>
> (quoted in Jefferys 1986: 113)

Chrysler was a somewhat special case, but by no means an isolated incident. At US Steel, for example, fractional bargaining was also widespread. Indeed, the fruits of extra-contractual fractional bargaining received quasi-legal protection in steel through a "past practice" clause in bargaining agreements, which prevented management from altering existing shopfloor conditions unless required by technological developments. Fractional bargaining, in combination with the past practice provision, was an important source of rank-and-file control over working conditions at US Steel throughout the 1950s (Betheil 1978). And this was true despite the existence of a no-strike provision and "management's rights" clause in the contract.

The extent of fractional bargaining also varied across jobs in the plant. From his field investigations in the mid-1950s, Sayles (1958) characterized four types of industrial work groups—apathetic, erratic, strategic, and conservative—all so named for the kinds of behavior they exhibited. The apathetic and conservative work groups represented the very bottom and very top, respectively, of the occupational hierarchy. Skilled workers possess power in the shop by virtue of their superior knowledge of production. Custom and practice is a terrain over which management reluctantly abdicates authority to skilled workers. Unskilled workers generally possess little power of any kind.

Semi-skilled workers composed the majority of workers in the erratic and strategic work groups. These groups struggled for control of shopfloor custom and practice with management in day-to-day encounters and in ways typically associated with fractional bargaining. Each group exercised shopfloor power in a different way, though. Erratic work groups—characterized by small assembly or work crew forms of production involving dirty, dangerous, and physically tedious tasks—were more prone to spontaneous outbursts of discontent. (An example is wet sanders in the auto industry.) Strategic groups on the other hand—characterized by individual production jobs involving self-pacing and critical judgment factors—were more enthusiastically engaged in strategic battles over custom and practice and manipulation of the grievance machinery. (Examples are welders, metal polishers in autos, or wire drawers in steel.)

Through fractional bargaining, unionized workers were able both to win new improvements in working conditions in the immediate postwar period and to protect past gains. This form of shopfloor power served as an adequate mechanism for policing those few contractual clauses covering shopfloor conditions, as well as being an important source of rank-and-file empowerment in influencing the many shopfloor conditions that were largely noncontractually determined.

The fruits of fractional bargaining were not limited to workers. For foremen, who were charged with the unfulfilling task of carrying out production standards or operating procedures concocted by people far removed from the realities of day-to-day production, fractional bargaining was a way for production to move forward without undue bureaucratic restriction. Moreover, workers who felt they had some control over their shopfloor conditions were more inclined to cooperate with management, and stewards who had garnered noncontractual shopfloor benefits for workers were to a certain extent beholden to foremen, and therefore willing to concede on other matters.

One of the big advantages for management was the greater flexibility that was accomplished: successful stewards could convince workers to take on an extra job task that was not part of their formal job description, to change shifts, work overtime, or even change jobs temporarily, as production orders varied. Successful stewards, and empowered workers, were even willing to entertain management's introduction of new work methods and machinery, confident that they would be able to protect shopfloor conditions during the transition. Thus, despite the possible inefficiencies that stemmed from a slower pace, or more careful attention to safety, there were also significant efficiencies in terms of enhanced cooperation with management.

Perhaps the productive inefficiencies exceeded the cooperative efficiencies, so that, even ignoring the wage and benefit increases garnered by unions over this period, fractional bargaining would have been viewed by employers as inferior to those systems of shopfloor governance—the "drive system" or company unions—that preceded it. But the earlier systems were no longer an option. And when workers' increased satisfaction is factored in, the comparison to older systems may well leave fractional bargaining as the superior form of shopfloor governance from society's point of view. Even Kuhn, who was no great fan of fractional bargaining, was willing to concede:

> Fractional bargaining, even when it undermines the decisions and poli-
> cies of higher management and the union officers, permits members of
> the work group to act more as men and less as machines. If there are
> sacrifices involved in output, efficiency, and use of resources, they may
> be worthwhile losses for the gain enjoyed, if one notes both carefully.
>
> (Kuhn 1961: 82)

Fractional bargaining existed as a system of informal shopfloor arrange-ments that granted significant power to workers in controlling shopfloor conditions. It took the form of a process of strategic grievance filing, shopfloor slowdowns, and occasional shutdowns, through which semi-skilled workers were able, with the support of shop stewards, to bring their power to bear on the conditions of production in the labor process. It operated largely outside the orderly confines of the contract-and-grieve approach to industrial rela-tions, whereby collective-bargaining agreements would contain a list of labor's sole rights in production, and the grievance procedure would be the sole mechanism for upholding those rights. Beginning in the mid-1950s, the system of fractional bargaining came under increasing attack by employers.

THE TRANSITION TO SHOPFLOOR CONTRACTUALISM

The reality of shopfloor governance did not initially accord with the mutual understanding arrived at between employers, union leaders, and the govern-ment during the immediate postwar period. This understanding nonetheless came to exert an important influence on developments in shopfloor relations during the late 1950s and early 1960s. It was during this period that employers began to make inroads in the area of custom and practice, reducing the scope of informal decision making and subjecting what remained to greater scrutiny by centralized authorities. At the same time, the government began to refashion bits and pieces of labor law, clarifying the bounds of collective bargaining and the scope of managerial prerogative in production, and throwing greater weight behind the system of grievance arbitration for dispute resolution. And unions developed bureaucratic, centralized bargaining structures that helped to win sizeable monetary rewards for workers, but at the same time insulated the labor leadership even more from rank-and-file workers' shopfloor demands. The effect of these developments was the rise of a system consistent with the basic tenets of a contract-and-grieve approach to shopfloor governance—a system I refer to as "shopfloor contractualism."

To the extent that employers were aware of the reality of shopfloor practice during the 1940s and early 1950s, the uninterrupted postwar expansion in economic activity served to contain their frustration with the lack of managerial control in production. The brief downturn in economic activity in 1957–8, however, caused employers' contentment with stability in production to wane, and forced a greater concern with the issue of unit labor costs. Another contributing factor in employers' increased concern with labor productivity and costs was growing international competition from Japan and Europe, which was observable in several industries by this time.[4] The shift in emphasis, combined with the often bitter struggles with workers surrounding the introduction of automated technologies, convinced employers to make some changes in shopfloor governance (Strauss 1962). While the alterations to follow were highly touted as being directed at the restrictive work rules contained in labor agreements, Livernash (1962) argues convincingly that the real target was the noncontractual shopfloor practices that workers had been able to establish over the years following unionization.

Employers spent enormous energy in the late 1950s and early 1960s reducing workers' shopfloor power in the realm of custom and practice. Management strove to get things in writing, to rationalize procedures, and to eliminate "bootleg" agreements. Workers' manipulation of the grievance procedure was circumscribed by increasing the size of industrial relations departments, hiring young, recently educated industrial relations experts with an exceedingly legalistic approach to the resolution of shopfloor disputes, and schooling foremen and supervisors in the proper approach to dispute resolution (Strauss 1962). The grievance procedure was meant solely as a mechanism for addressing contentious interpretations and alleged violations of the collective-bargaining agreement.

The evolution of company policy at Goodyear during this period is illustrative. Throughout the immediate postwar years, Goodyear maintained a strong commitment to decentralized resolution of shopfloor grievances and—unlike such notables as Ford and US Steel—the opportunity for workers to grieve any issue regardless of whether or not it was governed by contract language. A 1945 "Management Training Instructor's Manual," for example, encouraged supervisors to resolve grievances rather than sending them along to higher steps in the grievance procedure.[5] A similar message was offered in an early 1950s training session for supervisors entitled "Handling Problems at the Source."

Company documents reveal the reasons for this commitment to the decentralized resolution of disputes. Because actors at higher stages in the grievance procedure were thought to be unaware of the context of the grievance, it was feared that they might look only to see whether the grievance represented a contract violation. Lower-level grievance resolution, on the other hand, was more likely to yield an "equitable solution" because decision makers at this level were better informed of past practices and informal agreements that bore on the situation.

In the minds of Goodyear management, as stated in company documents of the period, "equitable solutions" were the "key to productivity." For roughly the first decade following the war, supervisor training courses instilled in supervisors the basics of the "human relations" approach to management, in the hopes of ensuring equity in shopfloor dispute resolution. The "leader type supervisor" was contrasted in these training sessions with the "driver type supervisor"—the latter being a reference to the method of supervision exhibited by foremen under the drive system of the early twentieth century, an approach that the human-relations school strove to eliminate. The human-relations approach towards management emphasized getting the facts surrounding a grievance and understanding the workers' position in raising the grievance. An early 1950s manual for supervisors advised that "The Human Touch Is An Attitude" and listed ten rules the successful supervisor should follow in dealing with workers, among them being "hear him, understand him, motivate him, counsel him, honor him."

Around the mid-1950s, a different emphasis emerged in supervisor training courses. The 1955 "Factory Supervisional Training" manual introduced for the first time a lengthy section on "Contract Interpretation." Supervisors were counseled to study the collective-bargaining agreement and to make decisions on grievances in light of this agreement. Visual aids were utilized to make the point. In one such aid, a barrel is shown with a "worker problem" entering on one side and a "supervisor decision" emerging from the other side. Written on the barrel are the words "Facts," "Contract," and "Policy." Supervisors were being encouraged to "Go Through the Barrel" in rendering decisions on grievances.

The following page of the training manual offers a possible insight into why the new, more legalistic approach to grievance handling was being pushed by management. The depiction is of a worker with a paddle labeled "grievance" and a supervisor leaning over a barrel ready to be paddled. The caption, continuing with the earlier admonition to "Go Through the Barrel," reads "Or, You May Be Over It!" The page following this one spells out the "management's rights" clause in the labor agreement and informs supervisors that umpires (i.e., arbitrators) have sustained this clause as governing residual issues not formally spelled out in the contract.

The following year, the 1956 "Factory Supervisional Training" manual took up more formally management's concern with workers' ability to influence noncontractual issues. Supervisors were advised to pay careful attention to " 'side agreements' (written, oral, 'gentlemen's,' etc.)" and to "examine past practice carefully". These, of course, are precisely the areas where fractional bargaining was paying off the most for workers.

Goodyear's efforts to alter the behavior of shopfloor supervisors during this period were not an isolated incident. There was, in fact, widespread concern among companies at this time about the role of the foremen in shopfloor management. General Electric, for example, conducted a series of studies in the mid-1950s on the foreman's job. A 1957 study entitled "The Effective Manufacturing Foreman" came to the following conclusions: the "more effective" foremen "spent more hours with staff and service personnel," had fewer "contacts with their own employees," possessed "less total work experience," were younger in age, and were "recruited from functional service jobs rather than directly from hourly rated production jobs" .[6]

Goodyear followed a common course in its attempt to eliminate fractional bargaining during these years: retraining foremen and line management, for example, and increasing the prominence and prestige of the industrial relations department. However, Goodyear also stood out among many firms during this period in trying to retain decentralized decision making in shopfloor governance. Indeed, during the late 1950s many tire manufacturers, acknowledging the importance of noncontractual custom and practice in shopfloor decisions, instituted a system of shopfloor bargaining through the use of "mutual agreements"—extra-contractual agreements hammered out between foremen and shopfloor labor representatives governing custom and practice in production (Kuhn 1961: 174–6). While this gave greater legitimacy to shopfloor bargaining, it also allowed all agreements to be subject to the approval of industrial relations departments, and reduced the scope of such activity by allowing upper management to decide which issues could be mutually agreed upon in this decentralized fashion.

The attack on fractional bargaining at Chrysler's Dodge Main plant in Detroit reflects the more conventional "centralization-of-control" strategy in employers' shopfloor initiatives during this period (Jefferys 1986). Chrysler's dwindling share of the car market—from 23 percent of domestic production in 1951 to 13 percent in 1954—acted as the initial call to arms for management to eliminate pereceived productive inefficiencies. The recession of 1957–8 offered even further provocation. Management focussed on the custom and practice of work standards determination, and the removal of past fractional bargaining gains at the department or shop level.

Developments during the late 1950s were devoted to eradicating the informal system of shopfloor bargaining.

These efforts began with a systematic plant-wide study of operations instituted by a redesigned time study department with an enlarged staff. Based on the study results, management embarked on a strategy of eliminating the worst of the perceived production inefficiencies. Speedups were instituted for a number of operations. For example, the ten minute per hour relief period won by workers in the heavy press shop during World War II was reduced to three minutes. A new policy was begun in which printed warning "tickets" were issued to workers who failed to keep up with the faster pace.

Chrysler management then used the 1957 model change as an opportunity to alter production standards for many of the remaining jobs. One worker, commenting on the speedup in the body-in-white shop of the plant, stated:

> The working conditions at the Dodge Plant have been at its lowest ebb since our Union was organized. Work schedules have been changed, not only in 1957 but 1956 as well. Nearly every worker has been putting out approximately 40 percent more production than in the past years.
>
> (Jefferys 1986: 136)

When the UAW proved unwilling to back union locals in their efforts to prevent the shopfloor losses, management pushed for even more changes in the recession of 1958.

In January, the director of labor relations at the Dodge Main plant summoned the local union president to a meeting, at which it was announced that the plant would be closed for a two-week period, and that, upon reopening, recalled workers would face a new set of job tasks and new production standards, both to be determined by new procedures and methods. Job content would henceforth be the terrain of the Industrial Engineers' Department, far removed from the shopfloor. The new method of analysis began with the total work time available per hour—the number of workers times sixty minutes—and the content of jobs was determined so that the workers found themselves busy for as much of the time as possible while they were on the floor. The new standards called for a speedup in final assembly jobs ranging from 17 percent to 43 percent (Jefferys 1986: 137).

The task of reducing the ability of shopfloor workers and stewards to influence custom and practice with regard to the new standards remained to be accomplished. An opportunity arose in June of 1958 for management to act. When the 1955 Chrysler/UAW contract expired that year, workers continued to work without a contract as part of the International's strategy, in reaction to the existing recession, of waiting for the start of the 1959 model year to engage in negotiations. Chrysler used this as an opportunity to make inroads on some of the contractual guarantees possessed by workers. Management instructed the chief stewards that they were to work on their regular jobs (and thus not engage in union business) for at least six hours a day, thereby diminishing the power of stewards at the plant. Foremen and superintendents were then instructed to ignore "chief stewards when they try to settle a grievance without reducing it to writing" (quoted in Jefferys 1986: 140), thus reducing the possibility of fractional bargaining gains by ensuring that grievances would henceforth be kicked upstairs for perusal by members of the industrial relations department.

All that remained was to codify these changes in the new contract terms with the UAW. On the matter of managerial prerogative over production standards, the old language, which read simply that "Management agrees that in establishing rates of production it will make studies on the basis of fairness and equity consistent with the quality of workmanship, efficiency of operations and the reasonable working capacities of normal operators," was changed to read:

> Clause 1: Time study is a generally accepted method used by management as a basis for establishing rates of production and as a basis for determining the fairness of work loads
> Clause 2: The method of technique used by management for the establishing of rates of production is an exclusive right to management, and the Union does not have the right to challenge the method that management selects.
>
> (quoted in Jefferys 1986: 143)

On the matter of informal decision making of production standards on the shopfloor, the new agreement further reduced the power of stewards by altering the context within which they could act on such decisions. Before, stewards could get access to data on the determination of production standards by simply asking foremen for the information, and then presumably haggling with them about the "fairness" or "equity" of the determination. Now, stewards would be made to file a written grievance in order to access such information, thereby ensuring that members of the engineering and industrial relations departments would be privy to any changes in company-determined standards of production.

Events in the steel industry during this period reveal how adjustments within the management ranks affected workers' ability to influence shopfloor outcomes through informal channels in that industry. Numerous interviews with steelworkers over the decades following World War II led Hoerr to conclude that "many aspects of worklife began to deteriorate in the early 1960s" (1988: 296). A steelworker with almost forty years of experience in the steel mills recalled:

In the 1950s, it was a pleasant place to work. Grievances were minimal. . . . [and there was a] good relationship with your immediate supervisor. [In the early 1960s] the company brought in college grads as supervisors. . . . This helped destroy working relationships. In the early seventies, they began hiring attorneys to run personnel services. Most of them used it as a stepping-stone and became tough bastards. 'If you want something, arbitrate!' they told us.

(quoted in Hoerr 1988: 296–7)

The bureaucratization of management structures and a more legalistic approach to shopfloor governance was responsible for the worsened social relations in steel. Arguably of the greatest significance was the increased use—beginning as early as the mid-1950s in some plants—of college graduates to serve as line foremen as part of their training for upper-level management positions. Prior to this, most foreman came from the hourly ranks of production workers. In the recessionary pressures of the late 1950s, replacing seasoned, highly paid supervisors with inexperienced college graduates was pursued as a cost-cutting measure. In April of 1961, following two consecutive years of falling profits, US Steel terminated thousands of salaried employees and began replacing them with younger college graduates. Hoerr writes, "The Mon Valley lost a high proportion of the oldest, most experienced [and highest paid] supervisors" (1988: 308–9).

The newer recruits had little interest in production management. Moreover, they stepped into jobs where they were increasingly asked to follow directives from higher levels in the management bureaucracy, and to refrain from settling grievances on the floor for fear that precedents might be set which would come back to haunt the company in the future. Production workers were thus being driven increasingly harder by foremen and supervisors who knew little about the process of production and who, themselves, were granted increasingly little say in determining how the work was done. "By the 1970s," writes Hoerr, "demoralization pervaded the Monongahela Valley mills" (1988: 296).

In the electrical equipment industry during the late 1950s, dominant firms such as General Electric and Westinghouse began noting, in company documents and in negotiations with unions, their growing disadvantage in international markets due to labor cost differences with competitors. They warned that existing market share, and therefore employment levels, could not be maintained in the future without considerable increases in labor productivity. Restructuring efforts, begun somewhat earlier in the decade, thus took on a renewed urgency as the decade progressed (Schatz 1983: 234–7).

To management's mind the major impediment to increased productivity was the power of workers and shopfloor union representatives. Workers' shopfloor empowerment had emerged during the organizing drives and early

118

days of unionization, and had gathered strength and durability since that time as a means of influencing, among other things, the process by which piece rates were set. Informal work group norms regulating work effort and shop steward influence over time study procedures provided workers with a distinct advantage in shopfloor labor–management struggle over the wage-effort bargain, a major determinant of unit labor costs. The willingness of foremen and supervisors to haggle with stewards over rates also played into the hands of workers.

Several solutions to this problem were adopted by the dominant firms in the industry during this period. One was the introduction of the "unit manager" system, in which young, college-educated junior executives replaced general foremen and department superintendents in shopfloor management. Don Tormey, a United Electrical Workers' organizer during this period, described the impact of the unit manager on shopfloor governance:

> He got tough with the union. He began to delay the whole goddamned thing. Would not handle grievances or would just flop 'em off—not settle things. And with a constant speed-up . . . .Tighten up. Increased productivity. The whole goddamned thing. Make it a harder place in which to work.
>
> (quoted in Schatz 1983: 237)

A second solution was the elimination of piece work altogether in favor of a measured daywork system. Under measured daywork, workers were paid fairly high hourly wages but were required to meet stringent company-set standards for quantity and quality of work. Yet another solution emerged in the early 1960s when the industry witnessed a tremendous burst of techno-logical change in production and product innovation. Tape-controlled machine tools undercut the power of machinists and printed circuits elimi-nated the jobs of wiremen, solderers, and inspectors. Semiconductors were also introduced during this period.

Organizational changes were thus accompanied by technological develop-ments that may have served to reduce the shopfloor power of workers in manufacturing during this period. Technological change during the 1950s and 1960s came in the form of further increases in the degree and scope of mechanization and the introduction of automated technologies.[7] The litera-ture abounds with stories of the tension accompanying the introduction of automated technologies on shopfloors during these years. An employer's failure to notify workers in advance of the appearance of automated equip-ment could cause bitter struggles between labor and management. Automation often significantly altered the occupational structure of the plant, the skills required of workers, and therefore the wages they received. But part of the tension surrounding the spread of automation appears to have resulted from the important role automation played in allowing employers to reduce the power of informal work groups, and thereby appropriate greater control in production.

Although the effects of automation varied across industries, scholars of technological change point to a number of common features.[8] Automation's impact on the workplaces of the 1950s—in such industries as automobiles, (Faunce 1958), steel (Walker 1957), chemicals (Blauner 1964), and electric power (Mann and Hoffman 1960)—tended to: (1) decrease the amount of team production; (2) reduce the ease of communication between workers, as it tends to increase the distance between work stations; (3) diminish workers' control over the pace of production; (4) increase the intensity of supervision, since a breakdown at one job location could often put a halt to the entire production process; (5) lessen the amount of physical effort and increase the level of attention required; (6) decrease the differentiation of tasks, thereby leveling the occupational structure of the plant, reducing the number of job titles, and limiting the size of job ladders for promotion; and (7) blur the distinction between white-collar and blue-collar jobs, particularly in continuous-process technologies.

Features 1 and 2 reduced the opportunity for informal work group coordination, while feature 3 suggests a direct worsening of shopfloor conditions as a result of automation. The tendency for automation to reduce the possibilities for informal work group coordination may have been counteracted to some extent by features 6 and 7, while not all shopfloor conditions were made worse by the introduction of automated technologies, as suggested by feature 5. Feature 4 is ambiguous in its effect on work group power; greater supervision reduces the space for informal work group behavior, but more continuous, sequential processing of materials in production allows isolated work groups to disrupt production to a greater extent. Arguably, the sum total of these changes, however, was a diminution of informal work group contact and cooperation, and worsening shopfloor conditions, especially in the area of work pace.[9]

THE PERIOD OF SHOPFLOOR CONTRACTUALISM

The rise of shopfloor contractualism and decline of fractional bargaining across many manufacturing industries during this period represented the substitution of a contract-and-grieve approach to shopfloor governance for a decentralized form of labor–management shopfloor co-determination. The scope for informal custom and practice in shopfloor decision making was severely curtailed, as responsibility for shopfloor production decisions was moved from foremen and supervisors to higher chains of command in the management ranks. In some cases, shopfloor dispute resolution was intended to remain decentralized, with newly trained foremen and supervisors enforcing company policy and granting workers' influence only when contract language guaranteed it. However, the grievance procedure was formalized, with more disputes being written up and therefore scrutinized, if not resolved, by upper-level management. Every attempt was made to limit labor's shopfloor rights to those formally spelled out in collective-bargaining agreements.

Did these changes in the institutional arrangements of shopfloor governance spell the demise of labor's influence over shopfloor conditions? As workers began to experiment with the new arrangements, it became clear that the answer to this question hinged on the answers to a number of more specific questions. What kinds of shopfloor issues could be submitted to the formal grievance procedure? On what basis would a decision be rendered if contract language was silent on the issue? Could shopfloor conditions, which involve day-to-day production uncertainties, really be subject to contractual regulation? Would the union leadership be supportive of workers' efforts to negotiate over language to regulate shopfloor conditions?

Employers were of different minds during this period on the question of what kinds of issues could be submitted to the formal grievance process. Some took hardline positions, maintaining that worker grievances must refer to specific violations of the labor agreement by management. US Steel contracts with the United Steel Workers (USWA), for example, contained a clause stating that grievances must involve "the interpretation or application of, or compliance with, the provisions of [the collective bargaining] agreement."[10] Other employers maintained a more open-ended approach to the issue of grievability. Surveys of firms from the late 1950s (Derber *et al.* 1961: 88) found that roughly half of all surveyed employers allowed grievances to be filed on any issue, roughly 20 percent allowed only contractual issues to be submitted, and the remainder fell somewhere in between.

As a strong proponent of the contract-and-grieve approach to industrial relations in unionized settings—as the American form of "industrial democracy"—government agencies responsible for overseeing the postwar industrial relations system were forced to confront this question as well. If, in this approach to democracy in industry, the labor contract is viewed as the "legislative" mechanism by which the workplace is ruled, and the grievance/arbitration procedure is seen as the "judicial" mechanism (Klare 1981: 467), then what is to be done with issues not covered in the "legislation" itself? The NLRB's response, as Stone (1981: 1,549) points out, has never been very clear. The Supreme Court, on the other hand, has seemed to favor the grievance/arbitration mechanism for the resolution of all kinds of disputes. In its 1960 *Steelworkers' Trilogy*[11] decision, for example, the Court held that doubts about whether an issue was grievable should always be made in favor of grievability (Tomlins 1985: 321).

For workers in firms that maintained such an open-ended policy towards grievance handling, this approach only brought forth yet another question—namely, on what basis were decisions to be rendered if worker grievances did not make reference to alleged violations of the collective-bargaining agreement? Management could presumably reject out of hand such a grievance, arguing that there was no basis for a violation of any employer obligation. The grievance could then make its way through the various stages of the procedure, landing, if pursued this far by the union,[12] on the desk of an arbitrator. But, then, how would the arbitrator decide?

An arbitrator's decision on whether to hear a case, as well as the ruling on the matter if it is heard, typically appeals to the concerns of fairness and justice, certainly to past practice, but also to managerial prerogative. No less an authority than Sumner Slichter maintained that arbitrators generally view disputes over production issues such as job content (i.e., production standards and job tasks) as "a fundamental right of management" (Slichter et al. 1960: 253).

Assuming that worker grievances not covered by contract language were less likely to be resolved in workers' favor under a system of shopfloor contractualism, then could contract language be sufficiently expanded to cover the many shopfloor conditions of interest to workers? There were several limitations here. One is that it is virtually impossible to anticipate, and therefore regulate in advance through contract language, all future shopfloor eventualities. A second limitation is that it is extremely costly—in terms of time and energy—to jointly agree to and write contract language covering many of those shopfloor conditions that can be anticipated. Under fractional bargaining, shopfloor labor and management responded jointly and spontaneously to shopfloor eventualities as they arose. Under shopfloor contractualism, because of the limitations just mentioned, many of these eventualities would fall under the formal control of management.

Finally, for those shopfloor conditions that seemed most amenable to contractual regulation, there remained the question of union leaders' support for negotiations over such issues. Several obstacles stood in the way of this support. Over the postwar period there was a slow drift away from the principles of "social unionism" towards those of "business unionism," and this was accompanied by a narrower, more material definition of collective-bargaining goals (Lens 1959; Davis 1986). Shopfloor demands were associated with labor militancy, challenges to managerial prerogative in production, and even Communist commitments among shopfloor activists. The labor leadership had grown more conservative.

Union leaders had also become fairly well insulated from rank-and-file demands by this time, allowing them to act on their conservative beliefs with greater freedom. Some had come by this relative autonomy by virtue of long-standing internal rules, such as the inability of union members to directly elect their leaders. Developments during the years following the war aided in these efforts. The NLRB's "contract-bar" doctrine, for example— which protected an existing union from rivals for either the length of the contract period or for a "reasonable proportion" thereof—served to protect the leadership from rank-and-file demands by preventing workers from defecting to rival unions (Tomlins 1985: 322). The longer contract terms—from an average of one year in the early 1940s to three years and beyond later on—reinforced the insulation.

Unions had also become bureaucratic organizations over the postwar period, and thereby fairly impenetrable to the demands of their constituents. There were some very good strategic reasons for this. Because the postwar industrial relations system in the US required workers to negotiate privately most of the benefits they won from employers, centralization of union power at the level of the international, multi-plant and multi-employer bargaining agreements, and contract negotiations conducted by well-trained experts, became crucial factors in winning wage and fringe benefits demands from firms within an industry.[13]

However, these developments served one set of workers' interests at the expense of another. As the power to affect wages and fringe benefits grew, the ability to influence shopfloor conditions diminished (Weber 1967). The growing bureaucracy made communication between the rank and file and the union leadership more difficult; the centralization of power allowed labor leaders to be less than fully responsive to workers' shopfloor demands; and the need for coordinated bargaining across plants and firms left little room for local issues demands.

Developments in labor law during the late 1950s and early 1960s may also have served to dampen the success of workers' efforts to win contractual demands covering shopfloor conditions. Arguably the most important of these developments were attempts by the NLRB and the courts to spell out the hitherto vague line separating managerial prerogative from those "other working conditions" over which labor had been granted the right to bargain collectively by the 1935 Wagner Act.

This distinction—between managerial prerogative and bargainable work-place issues—was addressed by government authorities during this period by way of setting out "the proper subjects of collective bargaining." The central issue was whether employers were compelled to negotiate over *any* aspect of working conditions during collective bargaining rounds. The answer came, albeit in an extremely vague form, with the Supreme Court's distinction between mandatory and permissible subjects of bargaining. In the famous *Borg-Warner*[14] decision of 1958, The Supreme Court stipulated some things as "mandatory subjects" of bargaining and others as merely "permissible subjects." In the case of mandatory subjects, the parties were duty-bound to discuss such issues to impasse in negotiations, at which point each party had the right to act unilaterally (in labor's case, strike) in order to bring its influence to bear on the issue. However, no such obligation to negotiate, and no protection for labor should it choose to strike, existed for permissible subjects.

In a series of Board rulings in 1961 and 1962, and upheld by the Supreme Court in its *Fireboard*[15] decision, there was an attempt to set out

123

"meaningful limits." Certain issues, such as technological change, were deemed to be clearly within the realm of managerial prerogative, and thus considered to be only permissible subjects of bargaining. The Supreme Court's wording is instructive:

> Nothing the Court holds today should be understood as imposing a duty to bargain collectively regarding such managerial decisions, which lie at the core of entrepreneurial control. Decisions concerning the commitment of investment capital and the basic scope of the enterprise are not in themselves primarily about conditions of employment. . . . If, as I think clear, the purpose of (Section) 8(d) is to describe a limited area subject to the duty of collective bargaining, those management decisions which are fundamental to the basic direction of a corporate enterprise . . . should be excluded from that area.
>
> (quoted in Klare 1978: 320–1, footnote 198)

Permissible subjects of bargaining were not made illegal by these rulings, and so a very powerful union could force an employer to discuss such matters during contract negotiations. Moreover, many of the substantive shopfloor conditions of concern to workers fell within the vague categorical confines of mandatory subjects of bargaining. Thus, the effect of the mandatory/permissible distinction was probably largely ideological; it reinforced the conservatism of labor leaders on control issues in production, and bolstered employers' sense of righteousness in reestablishing managerial prerogatives over shopfloor production.[16]

These various limitations of the system of shopfloor contractualism for the attainment of workers' shopfloor goals became apparent shortly after the new institutional arrangements went into effect. In 1961, strikes at GM and Ford took place after the master agreements had been negotiated. While wages, hours, and benefits had been agreed upon, workers demanded time to address plant-level shopfloor concerns in local contract negotiations. GM alone was presented with 19,000 local demands during this strike (Livernash 1967).

During the 1965 negotiations in steel, demands on local issues, most of which dealt with conditions on the shopfloor, posed a significant threat to the *status quo* negotiating structure. Committees were set up at the plant and company levels, in addition to the normal industry-level bargaining committee, to conduct negotiations. As a result of these initiatives, more issues concerning working conditions began to appear in contracts, but as Livernash (1967) makes clear, contract language was no substitute for the resolution of workers' shopfloor concerns through decentralized forms of power and decision making on the shopfloor.

The formal, bureaucratized grievance procedure that emerged during these years was no substitute either. By the mid-1960s the number of filed grievances in many major manufacturing firms had shot up dramatically, and those going unresolved were beginning to mount. At GM, for example, the number of written grievances per 100 blue-collar workers rose from 27.0 in 1960 to 71.9 in 1973 (Lichtenstein 1986: 135). Lichtenstein notes that the number of unresolved local grievances that were raised during contract negotiations at GM amounted to 11,600 in 1958, but grew to 39,000 by 1970 (1985: 370).

Herding found from interviews with management and labor representatives in a steel plant in the late 1960s that "the (grievance) load has increased at a rapid pace in about a decade. . . . The speed of the procedure is 'definitely stalled'" (Herding 1972: 188). Arbitration cases, for example, took an average of sixteen months to resolve. In this same plant during the period 1948–52, an average of 6 percent of all grievances appealed to arbitration were still pending at year's end. By 1967, this number was 49 percent!

Formalization of the process of shopfloor dispute resolution instituted an "obey now, grieve later" aspect to shopfloor practice, granting management the unilateral right of immediate shopfloor governance, subject only to later review by grievance committees or arbitrators. Apart from the frustration of waiting for a resolution, workers realized that the longer the period before a dispute was resolved, the greater was management's prerogative in production.

Livernash, noting the emerging crisis in the grievance procedure during this period, wrote "This invites the query whether, in addition to a traditional negotiation process far removed from the individual employee, the grievance procedures may have become so formal and precedent-laden as not to be a sufficient outlet for employee complaints," (1967: 44). And Ulman stated, "Either the grievance procedure must be made to work effectively or the local unions must be given the right to strike on important grievances after reasonable efforts to resolve them peacefully have been frustrated by management" (quoted in Livernash 1967: 44).

In March of 1964, delegates to the UAW convention complained bitterly about the unbearable working conditions in the auto industry. Of particular concern was the issue of speedups, which, according to delegates, had caused nervous disorders in some workers who were unable to keep up with the increased pace of production (Mkrtchian 1973: 45). GM seemed to be the biggest offender, and the International, perhaps recalling the events of the 1961 bargaining round, decided to press the issue firmly in upcoming contract negotiations.

Walter Reuther negotiated a sizeable increase in the grievance handling time of committeemen, thinking this would be sufficient to address workers' concerns. However, an eleven-member negotiating committee from the GM locals—an institutional structure which itself had grown out of the workers' 1961 demands—turned down the contract, and after a week-long national strike in which national issues were finally resolved, there ensued a militant, 31-day strike over local issues by the rank and file.

B. J. Widick wrote in *The Nation*: "The nearly six-week shutdown . . . of

General Motors must be described as the most prolonged and biggest 'wildcat' strike against American industry since the sit-downs of the turbulent thirties" (349). During local negotiations at GM Ternstedt in Flint, Michigan, 121 demands on local issues were brought by UAW Local 326 to the bargaining table, requiring fifty-one meetings and over a month's time to resolve (Lichtenstein 1986: 131). Only a short time later, many Ford locals went out for improved working conditions.[17]

The crisis associated with the system of shopfloor contractualism escalated as the decade of the 1960s drew to a close. The percentage of strikes over working conditions rose from an average of roughly 16 percent in the period 1953–60 to almost 30 percent in the period 1968–73. The percentage of wildcat strikes, over half of which are typically attributable to working conditions issues, also increased over the 1960s, from 32 percent of all strikes between 1961 and 1967 to 40 percent between 1968 and 1973 (Naples 1981: 38).[18] Worker survey results from the 1960s and early 1970s, the most famous of which is the *Work in America* report (1972), document the rising workplace discontent during this period.

Workers' shopfloor discontent with the new system of shopfloor contractualism penetrated popular consciousness with the now famous 1972 sit-down strike at the GM assembly plant in Lordstown, Ohio. In the late summer of 1971, GM introduced minor technological changes, brought in an entirely new management team, and altered the model design of the automobile in order to bolster the plant's productivity. The problem, it seems, was that informal agreements between workers and supervisors had left far too much power in the hands of the rank and file. As a member of the new management team put it, "When you have a plant over many years, it is very possible to have developed habits that aren't right: they are outside the (collective bargaining) agreement" (quoted in Aronowitz 1973: 42). The reorganization effectively eliminated these "habits," allowing production at its peak to reach 100 cars an hour (one every thirty-six seconds)—roughly a 60 percent increase in productivity compared with former shopfloor practice.

The sit-down at Lordstown was reminiscent of the labor actions of the late 1930s. While the organized workers of the 1960s possessed good wages, decent fringe benefits, the elimination of arbitrary treatment in hiring, firing, promotion, and layoffs by management, and a modicum of employment security, many shopfloor conditions were no better than they had been at the beginning of World War II. The system of shopfloor governance had apparently come full circle; workers' shopfloor influence had become so severely circumscribed that there existed no means by which workers could make shopfloor demands except those of an earlier time.

POSTWAR SHOPFLOOR GOVERNANCE AND SHOPFLOOR OUTCOMES

Did these postwar developments in shopfloor governance have an impact on such shopfloor outcomes as the trajectory of industrial safety and productivity growth over the postwar years? Several studies suggest that they did. For example, Derber *et al.* (1961) surveyed company managers and local labor officials in forty-one manufacturing firms in 1955 and then again in 1959. Their survey results suggest that unionized workers faced significant reductions in their ability to participate in decisions concerning the content of jobs and the safety rules governing the plant. Herding's (1972) more comprehensive study of the postwar period, including the period of the 1960s, also presents evidence of a decrease in the effectiveness with which rank-and-file workers were able to protect shopfloor conditions beginning in the late 1950s.

Aggregate measures of safety and productivity growth over the period seem to accord with these findings. During the period of fractional bargaining, from roughly the mid-1940s to the late 1950s, workers' shopfloor power was significant, and yet there was much cooperation between shopfloor labor and management because workers' empowerment ensured the legitimacy of managerial authority. Thus, we would expect shopfloor conditions such as safety to have improved, and productivity growth, bolstered by significant labor–management cooperation but also impeded somewhat by workers' enhanced control over work pace, to have been moderate. The injury frequency rate in manufacturing (the number of lost workday accidents per million labor hours) declined from 17.2 in 1948 to 11.4 in 1957.[19] The average annual growth rate in manufacturing labor productivity for the period 1948–60 was 2.43 percent.[20]

Between the late 1950s and the mid-1960s, the system of shopfloor contractualism was becoming firmly rooted in many major manufacturing industries. Workers' shopfloor power was becoming severely compromised, but workers sought refuge in the contract-and-grieve approach to shopfloor governance, introducing new shopfloor language into local agreements where possible and making greater use of the formal grievance procedure. The inadequacies of the system of shopfloor contractualism for regulating shopfloor conditions were not yet fully apparent to workers. We might expect productivity growth to have been more rapid, due to the enhanced power of management in production, and injury rate improvements to have moderated. Labor productivity growth was 3.9 percent annually from 1960 to 1966, while injury rates rose rather modestly from 11.4 in 1957 to 11.9 in 1963.

By the mid-1960s, the system of shopfloor contractualism was beginning to reveal its inadequacies as a mechanism for worker influence over shopfloor conditions. While employers were consolidating their shopfloor gains,

127

managerial authority was being openly contested, and slowdowns, quickie strikes, and a work-to-rule approach by labor had emerged in shopfloor production. We might expect workplace safety to have suffered further losses, and labor productivity growth to have been at its lowest level of the postwar years. And indeed annual labor productivity growth slowed rather abruptly, to 1.7 percent for the period 1966–9, while injury rates rose from 11.9 in 1963 to 15.2 in 1970.

The aggregate trends in postwar labor productivity and injury rates correspond rather nicely with what we might expect given changes in the institutional arrangements of shopfloor governance over these years. However, even better evidence can be marshaled in support of an association between institutional change in systems of shopfloor governance and changing shopfloor outcomes. The next two sections offer a more careful analysis of the postwar trajectory of injury frequency rates and productivity growth in manufacturing.

Postwar injury rates in manufacturing

The injury frequency rate in manufacturing declined precipitously in the decade-and-a-half following World War II, but rose with almost equal force during the 1960s (see Figure 5.1). By the late 1960s accident rates in manufacturing as a whole were only marginally better than they had been in the late 1940s, while in some industries injury rates reached a postwar high. The 1970 injury rates in meat packing and tires and inner tubes, for example, were higher than they had been in 1948.

There are surprisingly few studies of this rather alarming trajectory in manufacturing injury rates. Moreover, those few studies that do exist find little support for the view that there was a structural deterioration in manufacturing workplace safety during the 1960s (Smith 1973; Chelius 1977). The rising injury rates of the 1960s are found to result instead from two transitory phenomena: the rapid economic activity of the decade and demographic shifts in the manufacturing labor force.

The available evidence offers strong support for the claim that injury rates are procyclical—that is, that they vary positively with the level of economic activity. However, it is not exactly clear why this is the case. Rapid economic activity may place increased pressure on the pace of production and require that workers put in longer hours, both of which might lead to an increase in workplace accidents. Labor turnover among the existing workforce and the rate of new hires also tend to rise during periods of rapid economic activity, which means that there are likely to be more workers who are less experienced in their jobs, and therefore perhaps more accidents. Smith (1973) found that after controlling for the level of economic activity, there was no discernible linear trend—either upward or downward—in manufacturing injury rates for the postwar period to 1970.

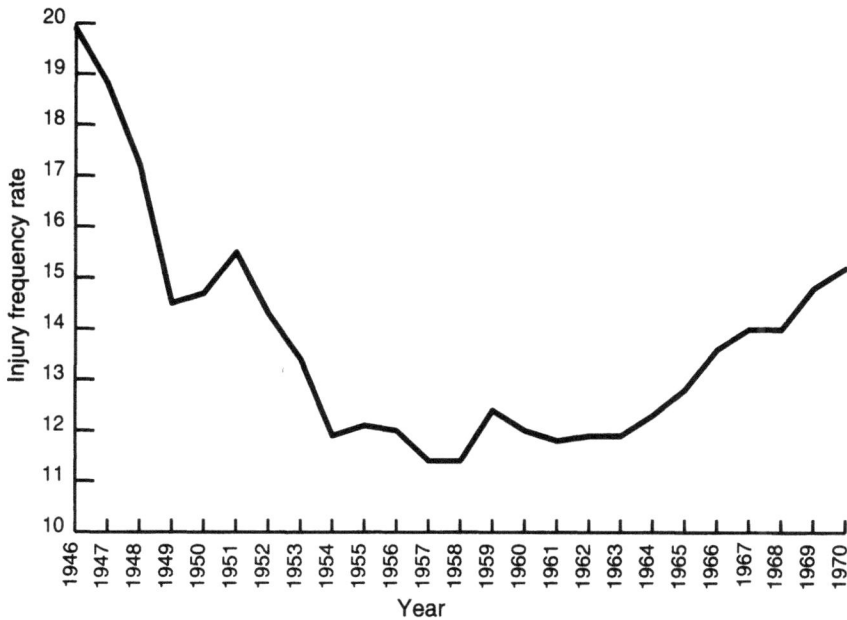

Figure 5.1 US manufacturing injury frequency rates, 1946–70
Source: See data appendix

Chelius (1977) took a different approach, but reached similar conclusions. He examined changes in the determinants of injury rates over the postwar period to 1970, and then tested for whether there was any evidence of a structural shift in the way in which these determinants influenced injury rates in the early postwar years (to 1964) compared to the later years (1964 to 1970). His determinants were the accession rate—which captures labor turnover including new hires—and the percent of the manufacturing work-force composed of younger workers. Younger workers are less experienced and perhaps more carefree, and are therefore hypothesized to be more prone to accidents. Both accessions, due to rapid economic activity, and the percent of younger workers, due to demographic changes in the working-age popu-lation, rose during the 1960s, thereby driving up injury rates. However, Chelius found no evidence to indicate that the shopfloors in manufacturing had become inherently less safe during the 1960s; if these two determinants of injury rates had not risen, safety would not have deteriorated.

Thus, the existing literature, sparse as it is, leads to the conclusion that the rise in injury rates in the 1960s can be accounted for by rapid economic activity and changing demographic features of the workforce. A recent reex-amination of this evidence, however, suggests that such a conclusion is premature (Fairris forthcoming). The new evidence reveals that even after

129

controlling for the level of economic activity, there is a very strong nonlinear, U-shaped trend in manufacturing injury rates over the postwar period, with 1960 being the year during which injuries began their postwar rise. Moreover, other determinants—in addition to accessions and the percentage of younger workers—seem to account for changes in postwar injury rates, and when these are added to the injury rate analysis, there is a clear structural shift in the underlying safety of manufacturing enterprises in the 1960s.[21]

In addition to a reexamination of the existing evidence on the aggregate injury rate in postwar manufacturing, Fairris (forthcoming) analyzed the trajectory of injury rates, controlling for the level of industry economic activity, in the various industries that compose the manufacturing sector. Eleven of the seventeen manufacturing industry groups for which data were available displayed a U-shaped trend in injury rates. These are listed in Table 5.1, along with the year during which the trajectory of injury rates in each industry reached its minimum. Note that injury rates bottomed out during the first four years of the 1960s in eight of the eleven industries, and for five of these industries the minimum occurred in either 1960 or 1961. The U-shaped pattern in postwar injury rate movements was apparently widespread across manufacturing industries.

If, contrary to conventional wisdom, the rapid economic activity and changing demographics of the labor force cannot fully account for the rising injury rates in manufacturing during the 1960s, what can? The introduction of automated technologies in many manufacturing firms during this period may account for some of the rise in injury rates, but neither German nor Japanese manufacturing witnessed a decline in shopfloor safety during these years (see Figure 5.2) and yet automation was sweeping manufacturing processes in these countries as well.[22] Thus, we are left searching for an explanation that is unique to the US and involves neither cyclical nor demo-graphic forces.

The explanation offered here, of course, is that the institutional arrange-ments of shopfloor governance, and especially the extent to which these arrangements promote or discourage expressions of shopfloor power by workers, can account for the decline as well as the rise in injury rates over the postwar period. Note, to begin with, that the timing appears to be right—and not just for the aggregate manufacturing sector, but for specific industries as well. For example, the case studies suggest that institutional changes in shopfloor governance were introduced in rubber during the mid-1950s; in autos, during the late 1950s; and in steel, during the early 1960s. Table 5.1 shows that injury rates reached their minimums in rubber, trans-portation equipment, and primary metals in 1955, 1960, and 1961 respectively.

Marshaling evidence of a statistical sort in support of our hypothesis would seem to require development of a measure of shopfloor power with which changes in postwar injury rates might then be associated. Toward that

Table 5.1 Industry injury experience

Manufacturing industry	*Year of injury rate minimum*
Apparel	1964
Chemicals	1962
Fabricated metals	1961
Furniture and fixtures	1961
Instruments	1963
Leather	1956
Lumber	1968
Primary metals	1961
Rubber	1955
Stone, clay and glass	1960
Transportation equipment	1960

Note: food, paper, petroleum, printing and publishing, textiles, and tobacco did not
display a U-shaped pattern in injury frequency rates during this period.

end, I surveyed by mail twenty labor historians and industrial relations
scholars and asked them to rank—on a scale of one (indicating low) to four
(indicating high)—the shopfloor power of *unionized* workers during the
postwar period to 1960 in each of nineteen two-digit manufacturing industries.[23] (I shall control for the percent unionized across industry workforces
in the empirical results that follow, but note that shopfloor power varied
widely across unionized establishments during this period—thus, a unionization measure would be a poor proxy for shopfloor power.) The responses
were averaged and a continuous measure of industry shopfloor power,
varying between one and four, was developed for each industry.[24] A ranking
of the industries by this measure appears in Table 5.2.

In row 1 of Table 5.3 the average annual percentage change in industry
injury rates over the immediate postwar period is regressed on the extent of
collective-bargaining coverage in the industry labor force in 1953 and the
measure of worker shopfloor power.[25] The results suggest that injury rates
declined the most in those industries where workers possessed the greatest
shopfloor power, and this is consistent with our hypothesis. The percent of
union organization, however, was not a significant determinant of the
interindustry movement in injury rates during this period.

The rise in the manufacturing injury rate beginning around 1960 is
explained, in part, by a reversal in the injury rate trajectories of a significant
number of manufacturing industries, from falling injuries during the

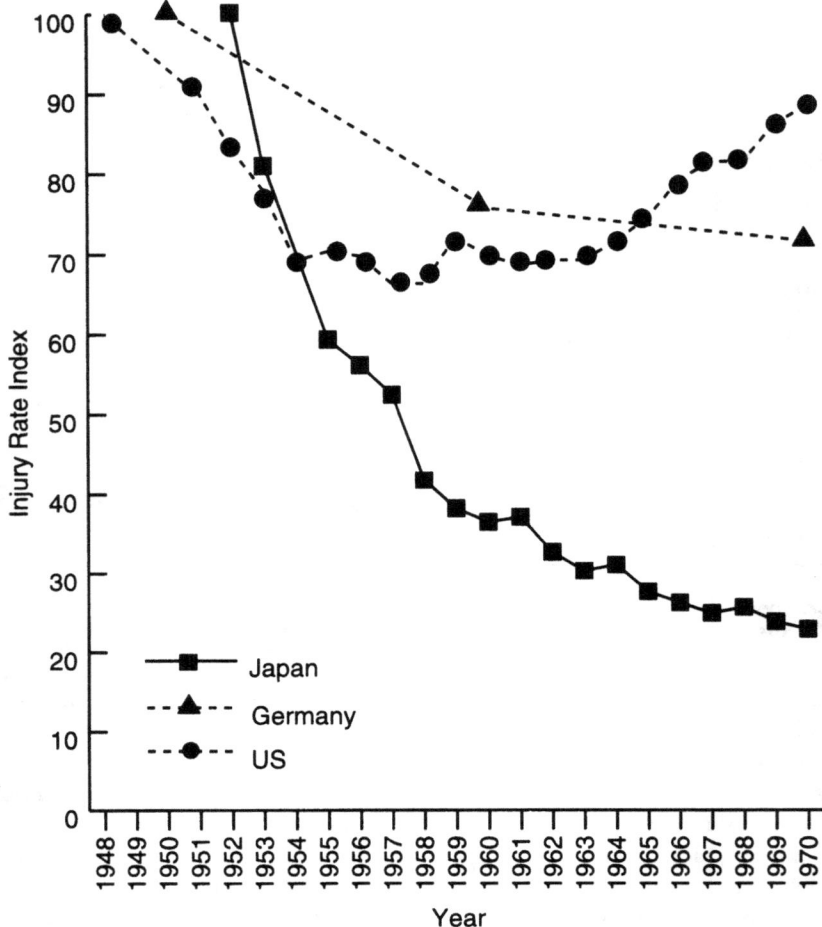

Figure 5.2 The trajectory of postwar injury rates: Germany, Japan and the US
Source: See data appendix

immediate postwar period to rising injuries during the 1960s. Can the institutional changes in shopfloor governance discussed above account for this reversal of fortunes in postwar industry injury rates? While our survey responses offer no information on the changing shopfloor power of workers across industries during this period, we conjecture that such changes were more likely to occur in industries where workers possessed significant shopfloor power in the immediate postwar years.[26]

Thus, as an indirect test of the hypothesis that institutional changes in shopfloor governance were responsible for the trajectory of postwar injury rates in manufacturing, we explore the relationship between the existence of a U-shaped pattern in industry injury rate experience and the industry-level

Table 5.2 A ranking of worker shopfloor power by
manufacturing industry, 1945–1960

High

          Machinery

          Printing and publishing

          Rubber

          Primary metals

          Transportation

          Electrical

          Instruments

          Stone, clay and glass

          Petroleum

          Chemicals

          Lumber

          Fabricated metals

          Leather

          Paper

          Furniture

          Apparel

          Food

          Textiles

          Tobacco

Low

---

measure of worker shopfloor power.[27] In particular, we use the results from
Table 5.1 to create a dichotomous measure of industry injury rate experience—equally 1 if the industry possessed a U-shaped pattern, and 0
otherwise—and related this to the survey measure of worker shopfloor power
by industry. The results appear in row 2 of Table 5.3.[28] They suggest that
while the extent of industry collective-bargaining coverage in 1958 had no
significant bearing on the tendency towards a reversal of industry injury rate
performance, the degree of shopfloor power across industries was a positive
and significant determinant of the existence of a U-shaped pattern in
industry injury rates.[29]

Table 5.3   Suggestive evidence on the relationship between shopfloor power, the trajectory of postwar injury rates and total factory productivity growth

Regression analysis

	Dependent variables	Independent variables			N
		% Union	Shopfloor power	Injuries$_{1948-1960}$	
(1)	Injuries$_{1948-1960}$	+	−*		19
(2)	U-shape$_{0/1}$	+	+**		17
(3)	Productivity$_{1948-1960}$	+	−		19
(4)	Productivity$_{1948-1960}$	+	−*	−	19
(5)	Prod. inc.$_{0/1}$	−*	+**		19
(6)	Prod. dec.$_{0/1}$	−**	+**		19

*   significant at the 0.10 level (one-tailed test)
**   significant at the 0.05 level (one-tailed test)

Postwar productivity growth in manufacturing

The aspect of postwar productivity growth that has attracted the most attention in the literature is the so-called "productivity slowdown" that began around the mid-1960s.[30] One can draw two relatively uncontroversial conclusions from this literature. First, the productivity slowdown contained two distinct phases, the first occurring in the late 1960s and the second occurring in the early-to-mid 1970s. Second, formal empirical models which relate productivity growth to factors such as growth in the capital/labor ratio, the rate of investment in research and development, shifts in the composition of output, the rise in energy prices, and increased government regulation, are much better at explaining the second phase of the slowdown than the first. At best, roughly 50 percent of the slowdown in productivity growth of the late 1960s can be explained by these factors. What accounts for the remainder?

There exists published evidence to suggest that this slowdown may be attributed to the emerging crisis in shopfloor governance during the late 1960s. Cross-sectional analyses of firm and plant-level data, for example, reveal that industrial relations indicators such as the number of grievances filed, the number of unresolved grievances, and the number of unauthorized work stoppages are important determinants of productivity differences across comparable plants (Norsworthy and Zabala 1985; Ichniowski 1986).

Those plants possessing high levels of these indicators tend to have lower levels of labor productivity. All of these indicators were on the rise during the late 1960s—signs of the systemic failure of the contract-and-grieve approach to shopfloor governance—thereby suggesting a likely decline in the rate of productivity growth.

Recent time-series analysis of productivity growth over the postwar period also suggests that labor–management conflict played an important role in the productivity slowdown of the post-1965 period (Weisskopf *et al.*1983; Naples 1988). Naples' regression results analyzing the determinants of the annual growth rate in manufacturing labor productivity from 1951 to 1980 are reproduced in Table 5.4. In addition to the variables more commonly associated with productivity growth—e.g., capacity utilization and the capital–labor ratio—Naples finds that the strike frequency, labor quit rate, accident rate, and a variable capturing a weighted sum of strikes and quits (the industrial conflict index) are all negatively and significantly associated with productivity growth. Moreover, the industrial conflict index and the accident rate explain a large percentage of the post-1965 slowdown in productivity growth in manufacturing.

The manufacturing productivity slowdown of the post-1965 period did

Table 5.4 Production-worker productivity growth in manufacturing, 1951–80[†]
(sign and significance reported)

	Dependent variable	Independent variables							N
		Capacity utilization 1951–80	Capital/ labor 1951–80	Energy/ labor 1951–80	Industrial conflict 1951–80	Strikes 1951–80	Quits 1951–80	Injuries 1951–80	
(1)	Productivity 1951–80	+	+***	+**		−***			30
(2)	Productivity 1951–80	+*	+***	+*			−***		30
(3)	Productivity 1951–80	+	+***	+**	−***				30
(4)	Productivity 1951–80	+	+***	+	−**			−***	30

*   significant at the 0.1 level (one-tailed test)
** significant at the 0.05 level (one-tailed test)
*** significant at the 0.01 level (one-tailed test)
†   Adapted from Naples (1988, 162)

not reflect the power of workers to reduce their intensity of labor effort, as was true, for example, during the late 1930s, when productivity growth slowed in those industries where workers had attained significant shopfloor empowerment. The productivity slowdown of the later period was based on worker disillusionment and despair, and stemmed from such things as rising rates of absenteeism, defects in production which caused the labor hours per product to rise, and a general "work-to-rule" approach to production by workers. These expressions of discontent remained long after the strikes and quits subsided. In the early 1980s a survey of management by McKersie and Klein (1985) found "employee motivation" to be among the most important constraints firms faced in trying to improve productivity.

Is there any evidence to suggest that the slowdown in productivity growth was specifically related to workers' frustration with their lost shopfloor power in the system of shopfloor contractualism? And, of equal interest, what is the relationship between productivity growth and shopfloor power in the period of fractional bargaining, and during the transition years of the early 1960s? In rows 3 through 6 of Table 5.3 we explore the impact of shopfloor power on rates of total factor productivity growth across industries during the postwar period.[31]

One of the most important shopfloor concerns of workers during the organizing drives of the 1930s was the pace of work, and enhanced control over work pace was one of the most important achievements associated with the attainment of workers' shopfloor power. We might, therefore, expect productivity growth in the immediate postwar years to have been relatively depressed in industries where workers' shopfloor power was the greatest, as appeared to be the case during the late 1930s. (See the empirical results of chapter 3.) However, the postwar system of fractional bargaining differed from the more disruptive model of the late 1930s; workers' control over shopfloor conditions had, by the postwar period, become institutionalized in the day-to-day flow of production, and labor's willingess, and indeed ability, to cooperate with shopfloor management had thereby increased.

The implications of these developments for a cross-sectional analysis of industry shopfloor power and productivity growth in the immediate postwar years is not clear. If we posit that shopfloor power is positively associated with the extent of both worker control in production and worker cooperation with management, the impact of these two ostensibly competing effects on shopfloor productivity is difficult to predict. The results presented in row 3 of Table 5.3 reveal a negative, but also statistically insignificant, relationship between productivity growth and shopfloor power in the first decade-and-a-half following World War II, consistent with a claim that the two competing effects of shopfloor power on productivity growth were offsetting during these years.

Workplace health and safety is one of the areas in which labor–management cooperation is most likely to have a significant impact. Improvements in health and safety might also enhance productive performance, as we saw with the empirical results on company unions in chapter 1. Therefore, holding constant the cooperative effect of injury rate changes on productivity growth might allow us better to isolate the conflictual effect of shopfloor power on productivity through the reduction in work pace, if such an effect exists. The results presented in row 4 offer some support for the existence of a conflictual effect. Holding constant the cooperative impact of safety improvements on productivity growth, workers' shopfloor power is associated with significantly slower growth in productivity. Overall, though, the results still lead us to conclude that there was no statistically significant effect of shopfloor power on interindustry productivity growth during the immediate postwar period.[32]

Next, we explore the relationship between the extent of workers' shopfloor power in the immediate postwar period and the likelihood that an industry's average annual rate of productivity growth was greater in the early 1960s compared to the earlier period. To the extent that fractional bargaining allowed workers to limit the work pace and other aspects of shopfloor production that promote productivity at the expense of worker satisfaction, the rise of shopfloor contractualism in the late 1950s and early 1960s in these industries should have led to a boost in productivity growth.

The results of row 5 reveal that while the percent of the labor force unionized was negatively associated with the likelihood of increased productivity growth during the early 1960s, the earlier extent of shopfloor power was positively and significantly associated with the likelihood that an industry would witness a relative improvement in productivity growth.[33] This lends suggestive evidence to the claim that the emergence of shopfloor contractualism reduced the ability of formerly empowered workers to regulate their pace of production, thereby leading to increased productivity growth during the early 1960s.

Finally, we explore the relationship between the extent of workers' shopfloor power across industries in the immediate postwar period and the productivity slowdown of the late 1960s. The results presented in row 6 explore the relationship between workers' shopfloor power and the likelihood of a decline in productivity growth rates in the late 1960s compared to the early 1960s. They reveal a statistically significant relationship between the early existence of shopfloor power and the likelihood of witnessing a slowdown in productivity growth in the late 1960s. This is consistent with the hypothesis that worker frustration with the system of shopfloor contractualism was most pronounced in those industries where workers had possessed the most shopfloor power in the immediate postwar period, and, because of this frustration, these industries were more likely to suffer relative declines in productivity growth in the late 1960s. In combination with the time series results reported in Table 5.4, these results offer suggestive evidence in support of the claim that the productivity slowdown of the late

137

1960s can be explained, at least in part, by the institutional change in shopfloor governance from fractional bargaining to shopfloor contractualism that took place roughly a decade earlier.

CONCLUSION

Contrary to conventional views of the period, the quarter of a decade following World War II did not witness a stable set of institutional arrangements for the joint regulation of shopfloor conditions by labor and management. For roughly half of the period there existed an informal system of shopfloor governance based on decentralized decision making and cooperative, face-to-face dispute resolution. During the second half a more formal system arose, one which was steeped in more centralized decision making, a strict adherence to contract language, and bureaucratic procedures for dispute resolution.

The informal system promoted a healthy degree of cooperation between shopfloor labor and management, but its sizeable role for custom and practice also played into the hands of shop stewards and powerful work groups, who were able, under such arrangements, to garner significant improvements in the shopfloor conditions of workers. In response to this aspect of the informal system, employers initiated the development of a more formal system of shopfloor governance, in which the scope for custom and practice was reduced and workers' rights were limited to those guaranteed by contract language. The immediate effects of this change were a boost in productivity and reduced safety in production, as workers lost their ability to control both work pace and work hazards.

If these consequences of the change in the institutional arrangements of shopfloor governance were part of employers' intentions, other consequences—the stalled grievance procedure, for example—were almost surely unintended. Moreover, employers failed to fully anticipate the intensity of workers' response to perceived injustice in the new distribution of shopfloor rewards and illegitimacy in management's new authority. As tension between shopfloor labor and management rose, cooperation and shopfloor productivity declined, causing a crisis in shopfloor governance with detrimental effects for the many stakeholders in American manufacturing.

6

POSTWAR COLLECTIVE-BARGAINING AGREEMENTS

Shopfloor contractualism was initiated by employers largely through changes in company policy, management structure, and personnel. The system made contract language the exhaustive statement of workers' rights on the shopfloor, and the formal grievance procedure the primary mechanism for dispute resolution. After only a few years of experience with this contract-and-grieve approach to shopfloor governance, workers realized its shortcomings. As shopfloor conditions deteriorated and unresolved grievances mounted, rising worker frustration made for tense labor–management relations.

A case study of collective-bargaining agreements over the postwar period offers useful insights into various features of the system of shopfloor contractualism. Changes in contract language and addenda attached to labor contracts shed light on both the rise of this system of shopfloor governance and the general state of labor–management relations—including, for example, workers' growing frustration with shopfloor contractualism in the late 1960s. In addition, an extensive study of local agreements over this period reveals the extent to which shopfloor conditions became subject to greater contractual regulation upon the emergence of the contract-and-grieve approach to shopfloor governance.[1]

Although most of management's success at eliminating workers' fractional bargaining power during the late 1950s and early 1960s can be attributed to changes in company policy, the new "get tough" policy of management is also reflected in changing contract language. One of the changes observed in a number of collective-bargaining agreements during this period relates to management's efforts to limit the precedent value of workers' illicit shopfloor gains. Another reflects management's attempt to eliminate workers' ability to use shopfloor actions to win improvements in other areas.

In steel, management's discomfort with existing shopfloor practice became so intense that the failure to reach an agreement with the union on this matter during contract negotiations in 1959 resulted in a strike that halted steel production for 116 days. Management's primary concern centered around the famous "past practice" clause in steelworker contracts, which prevented management from unilaterally altering shopfloor conditions unless necessitated by technological change. Workers had been able, for roughly two decades, to use this clause as a form of quasi-legal protection for shopfloor improvements won through informal shopfloor practices.

The steel companies were ultimately unsuccessful in removing the past practice provision from bargaining agreements, but following their defeat many companies fought for, and won, contract provisions that eliminated the precedent value of any dispute over local working conditions which was resolved prior to the arbitration stage of the grievance procedure.[2] This effectively undercut workers' ability to utilize the past practice clause as a guarantee of past shopfloor gains. In combination with company policy changes that restricted the realm of informal practices in dispute resolution, more disputes were resolved as formal grievances, and these would henceforth be formally recorded as changes without precedent value.

Steel was not the only industry where shopfloor victories won by workers through informal practices had come to be viewed as worker rights based on past practice. Past practice was a general consideration many arbitrators used in reaching decisions on matters in which contract language was either vague or nonexistent. Thus, similar attempts to reduce the precedent value of grievances resolved at lower stages of the grievance procedure took place in many other industries. In rubber, for example, language to this effect appeared for the first time in both Firestone and Goodyear contracts beginning in 1961.[3]

In autos, the concern was not only with the precedent value of shopfloor victories won through informal shopfloor practices, but also with the impact of a clause in auto contracts that prevented grievances over shopfloor issues—such as production standards and workplace health and safety—from being sent to arbitration. All arbitrable grievances in autos were covered by a blanket no-strike provision during the length of the contract; adjudication by a neutral third party served in lieu of strikes and lockouts as the accepted method for dispute resolution. Because shopfloor matters were not subject to this no-strike provision, workers possessed the right to strike to resolve grievances in these areas. Empowered workers could presumably use this right to make gains in job content and workplace safety. In addition, workers apparently used strike threats over such issues to achieve victories in other areas where conventional channels of interest representation proved unsuccessful.

The 1955 agreement between Ford and the UAW attempted to put a stop to such maneuvering through a "confinement of issues" clause:[4]

> It is expressly understood and agreed that no grievance, complaint, issue, or matter other than the strikeable issue involved will be discussed or negotiated in connection with disputes to which this Section [Local Working Conditions] is applicable (51).

Another common change in contract language during this period concerned the contractual right of union representatives to time study jobs as the basis for resolving disputes over production standards. Beginning with the 1955 Goodrich contract with the URW, the union was granted time study rights over any production standards grievance that had reached the step prior to arbitration.[5] The 1961 master agreement between the UPWA and Swift similarly granted the union the right to time study those jobs whose production standards had become the source of worker grievances.[6]

It is not immediately clear from the contract language what exactly was behind this development. After all, shop stewards had for over a decade possessed the informal right to take part in time study procedures. Indeed, with the support of informal work groups, stewards had been able to parlay this right into significant influence over production standards in many industries. The new contract language appears to represent a change of course, maintaining, or even expanding, the use of time study, but formalizing the process and perhaps changing the participants involved from lower-level management and union representatives to those higher up the chains of command.

We get a hint of the motivation for these changes from an addendum to the 1959 UPWA–Swift agreement.[7] The addendum takes the form of two letters, one from the company president and a response from the union president. The company president makes reference to the "output problem" (146) the company was experiencing. The union president acknowledges the problem and agrees to "advise all Local Unions covered by this Agreement that they should not encourage employees to initiate or participate in a reduction of production," adding that "the International Union does not approve of such practice and will discourage it as a matter of policy" (147). The company's proposed solution is interesting—namely, that a union representative be trained in time study techniques.

Collective-bargaining agreements thus reveal some of the efforts of employers beginning around the mid-1950s to clamp down on the informal shopfloor activities of workers in order to stem the tide of shopfloor victories made possible by fractional bargaining. By the early 1960s, the new contract-and-grieve approach to shopfloor governance was in place, resulting in growing demands by local unions to introduce shopfloor conditions into bargaining agreements in an effort to establish in contract language what had been lost in informal shopfloor power.

Evidence of the emergence of such demands can be found in a variety of collective-bargaining agreements during these years. "Relief time" appeared as a worker right for the first time in many union contracts during this period—for example, at Goodyear in 1957 and at Chrysler in 1961.[8] Language concerning specific safety issues also emerged for the first time during these years. The 1964 master agreement between Swift and the UPWA contained language stating that safety boots and shoes were to be supplied by the company for specific jobs.[9] In 1963, workers at the Goodyear plant in Akron, Ohio received contractual assurance that "no employee shall be required to work on any job in the plant with which he is unfamiliar until he has received adequate safety training" (98)[10]. There was even an attempt in autos, where the pace issue had always been the most pressing shopfloor concern, to achieve contractual protections against the speedup. The 1961 Chrysler agreement, for example, guaranteed that workers would not be required to make up a loss in production due to machine downtime or lack of materials, and recorded management's commitment that job tasks and line speed would be suitably altered to reflect new changes in auto design.[11]

A more systematic account of the rise of shopfloor contractual demands during this period can be gleaned from a detailed analysis of local contracts in the automobile industry. Archival documents housed at the Reuther Library at Wayne State University in Detroit allow us to track the evolution of local contract language during these years, as well as the demands presented by workers during local negotiations, regardless of whether they were eventually conceded to by management and adopted as part of the local agreement.

UAW Local 579 at General Motor's Central Foundry plant in Danville, Illinois, is an illustrative case.[12] Prior to the late 1950s, an average of 1 or 2 local demands was adopted per contract round in local agreements, while the average number of demands made by workers rarely exceeded single digits. In 1958, however, 30 local demands were made, 6 of which were adopted in contract language. From there, the numbers rise rather rapidly: in 1960, 65 demands were made and only 3 were conceded to by management; in 1961, demands and adoptions were 86 and 3 respectively; in 1964, 123 and 30; and in 1967, 200 and 50.

Most of these demands related to shopfloor conditions. There were requests for doors in the restroom stalls, better lighting in certain work areas, and better heating and ventilation. Of greater significance, though, were demands related to the pace of work and the safety of production.

Beginning in 1955, in a section of the local contract entitled "Speed Ups," contract language guaranteed that management would "not require employees to work beyond an established fair days' work" (64). Apparently, the vagueness of such a guarantee led to further attempts by workers later in the period to improve on this language. In 1961, workers demanded that a weight limitation of 65 pounds be placed on bags and containers to be carried by hand; management refused. In 1964, workers requested that management:

> stop forcing employees such as iron pourers, core setters and so forth to work over and above the normal standards. Where an infraction of this demand occurs the employees involved will be granted an extra relief period pro-rated on the length of time the violation existed. (83)

142

Management refused.

Safety demands also appeared during this period. In 1964, there was a request that at least two workers be placed on mechanized lines during operation for purposes of safety. During the same year workers demanded that the company refrain from using railroad cars seriously deteriorated by age and thus found to have become "a safety hazard" (58). In 1967, workers requested that management eliminate the practice of using less qualified switchmen on yard cranes because the practice jeopardized safety.

The case of Local 579 is not an isolated one. A similar story emerges at the famous Fisher Body plant in Flint, Michigan.[13] Local demands did not surface during local negotiations between UAW Local 581 and General Motors until the 1961 agreement, but they increased rather rapidly thereafter, from 4 in 1961, to 144 in 1964, and 216 in 1970. The shopfloor issues were similar to those raised at the Danville plant—such things as lighting, ventilation, safety, and pace. In 1964, for example, workers demanded that two workers always work together when the job task involved yard cranes, ladders, roof work, and the like. (The company refused, arguing that the current practice was safe.) Workers demanded that only trained drivers operate equipment. In 1967, the local union requested that, for safety purposes, all repair work during equipment breakdowns be performed by skilled trades workers.

It is unclear from the language of many of these demands whether they were attempts by workers to do something about long-standing grievances over the conditions of production, or requests meant to reestablish conditions recently made worse by management discretion. As part of local negotiations during 1961,[14] UAW Local 634 at the Harrison Radiator plant of General Motors in Buffalo, New York, recorded the following sentiment under the heading of "Speed-Up" in its local contract, thus leaving the very clear impression that, in the workers' minds at least, recent work pace demands were made in hopes of reestablishing old standards:

> The auto companies each day are trying to get greater production at less cost. We are not opposed to this greater output if it is the result of increased technological efficiency. We are opposed to greater output at the expense of increased physical energy on the part of the worker. (83)

Management's efforts to institute a system of shopfloor contractualism were expected to result in greater demands by workers for contract language governing shopfloor conditions. This is the nature of a contract-and-grieve approach to shopfloor governance. However, the crisis this system produced in procedures for dispute resolution was not foreseen. It seems clear from contract language during this period that most employers maintained every hope that the process of dispute resolution would remain somewhat decentralized, with foremen and supervisors making firm, but informed, judgments concerning the merits of a formal grievance in light of the language of the collective-bargaining agreement. Shopfloor managers do not appear to have been up to the task. Disputes were increasingly kicked upstairs to higher stages of the grievance process for adjudication by those with more legal minds, and the system eventually became extremely overburdened.

Signs of distress in the system of dispute resolution were recorded in many collective-bargaining agreements beginning in the early 1960s, while all-out crisis seems to have emerged in many instances by the mid-to-late 1960s. In rubber, for example, the 1963 master agreement between Goodyear and the URW imposed limits on the amount of time that could elapse before a hearing was held on a grievance submitted for adjudication by the umpire.[15] The agreement between Local 2 and Goodyear's Akron plant[16] was more explicit about an emerging problem. In a memorandum appearing at the end of the contract entitled "Memorandum on Handling Grievances," there was mention of a discussion during local contract negotiations concerning "aspects of the local grievance problem." An understanding was reached whereby an "earnest and sincere" attempt would be made to "settle all grievances during the early steps of the grievance procedure," and committing the personnel manager to making the arrangements for joint union–management meetings to discuss solutions to these "local grievance problems" (123–4).

Very similar language appeared in steel contracts around this same time. The 1962 agreement between US Steel and the USW[17] obligated both parties to resolve grievances at the shopfloor level, and instituted measures to speed up the later stages of the grievance procedure by setting strict deadlines and allowing similar grievances to be lumped together for purposes of adjudication. In the 1965 agreement it was decided to undertake a thorough review of the grievance procedure.[18] A special committee was appointed to review and make recommendations to improve the functioning of the procedure in those cases where it was found to be "unsatisfactory in giving prompt attention to grievances" (25).

At Ford the 1961 agreement with the UAW contained a letter from the Vice President of Labor Relations to the Director of the National Ford Department in the union acknowledging the extensive discussions that had taken place during contract negotiations concerning problems with "effective functioning" of the grievance procedure.[19] The letter states that an agreement had been reached that company and union representatives would be dispatched to "areas needing improvement" in hopes that "an educational process" might help to foster solutions (323).

By the late 1960s, workers in many manufacturing industries were forced to conclude that the system of shopfloor contractualism was a shabby substitute for the more informal, empowered shopfloor voice they had possessed under fractional bargaining. With this realization, cooperative relations between workers and management underwent significant deterioration. A "work-to-rule" approach to production was adopted by labor, to which management responded with threats of discipline and discharge.

The evolution of contract language governing production standards in the agreements between Chrysler and the UAW from the early 1950s to the late 1960s captures quite nicely the changing shopfloor relations between labor and management in manufacturing over the postwar period. In the 1950 UAW–Chrysler agreement, production standards were referred to rather briefly in a separate section of the contract labeled "Rates of Production."[20] The gist of the language is that management agreed to set rates based on "fairness and equity" (15). (In reality, fairness and equity were not totally inaccurate descriptions of the basis for deciding shopfloor outcomes at Chrysler during this period, but more because of workers' shopfloor empowerment than because of management's earnest commitment to such principles.)

By 1961, the production standard's section of the agreement[21] was relabeled "Work Standards," and contained, for the first time, specific stipulations governing the intensity with which workers were to work, primarily through workers' rights to a certain amount of relief time.

In 1967, work standards language was relegated to a different section of the contract[22] labeled "Discharge and Discipline." Union and management no longer strove for "fairness and equity" in work standards, just in the method by which the company disciplined workers who failed to meet such standards:

> When a standard is not established, an employee who is following the prescribed method and using the tools provided in the proper manner and performing at a normal pace, will not be disciplined for failure to obtain an expected amount of production. . . . When imposing discipline for failure to follow a prescribed method or for failure to use the tools provided in a proper manner, an employee will be informed in writing in what respect he failed to follow the method or use the tools. Upon request, the Chief Steward will also be given the reason. (48–9)

As with much of manufacturing during the late 1960s, industrial relations in autos had become tense, labor–management cooperation had deteriorated, and discipline had become a problem. The relatively new system of shopfloor contractualism clearly was not working.

7

CONTEMPORARY EXPERIMENTS WITH NEW SYSTEMS OF SHOPFLOOR GOVERNANCE

Over the past few decades employers have engaged in numerous workplace experiments designed to alter various aspects of the postwar approach to industrial relations. The shopfloor, and shopfloor governance in particular, has been the focus of many of these experiments. Quality circles are an attempt to reduce the adversarialism between labor and management in production and to foster greater dialogue between shopfloor actors concerning ways to eliminate shopfloor inefficiencies. Teams of production workers and lower-level supervisors have been granted greater autonomy in shopfloor decision making in the hopes that they will use this autonomy to implement changes that improve productive efficiency. The focus is on enhanced labor–management cooperation and improved shopfloor labor productivity.

In part, these shopfloor experiments can be linked to such causal forces as technological change, innovations in the philosophy of human resource management, and increased international competition (e.g., Piore and Sabel 1984; Kochan *et al.* 1986). It is often claimed, for example, that new technologies and new approaches to the management of workers make it possible to move beyond the postwar system of industrial relations—a system which is seen as having promoted labor–management adversarialism, prevented flexibility in the allocation of labor, and hampered productivity and product quality. Increased international competition, on the other hand, has made it necessary that this postwar system be replaced.

However, there are deeper causal forces lying behind these employer-initiated shopfloor experiments, forces rooted in the crisis of shopfloor governance during the 1960s. Adversarialism and inflexibility were not universal features of postwar institutional arrangements. Adversarialism in labor–management relations became most problematic during the late 1960s, with the crisis of shopfloor contractualism. Inflexibility in the utilization of labor resources increased during this same period, as a result of workers' growing frustration with the ineffectiveness of the contract-and-grieve model of shopfloor governance. Workers' lack of devotion to productivity and product quality also originated during these years and resulted from the same source.

The contemporary efforts of employers to transform the institutional arrangements of shopfloor governance are devoted to addressing certain manifestations of the crisis of shopfloor contractualism, such as the overburdened grievance procedure, high rates of worker absenteeism, and the "work-to-rule" approach of workers in production. Quality circles and work teams represent a return of sorts to the decentralized framework of fractional bargaining, but with the power of informal work groups and shop stewards drastically curtailed. It strives to recapture the cooperative aspects of this earlier form of shopfloor governance, while at the same time altering those features of the earlier system that allowed workers to garner significant shopfloor improvements.

THE RECENT EMPLOYER EXPERIMENTS IN SHOPFLOOR GOVERNANCE

Quality of worklife committees and quality circles

Programs to foster greater communication between labor and management in production have grown rapidly during the past few decades. These programs have gone by different names, both across companies and over time, since their inception; earlier versions were often called quality of worklife (QWL) committees, while later versions are usually referred to as quality circles (QC). Joint-communication committees such as these typically consist of regular meetings of rank-and-file workers and management, taking place in company time, to solve problems encountered in production. The committees rarely have any direct power, as implementation of proposals almost always requires approval by upper-level management. Because labor law grants exclusive representation to duly elected bargaining agents, the existence of such committees in the unionized sector requires the tacit approval of unions.[1]

Quality of worklife committees in unionized manufacturing have their origin in programs begun in the 1960s and early 1970s by employers who were attempting to address the emerging crisis in the system of shopfloor contractualism—the jammed grievance procedure, workers' shopfloor discontent, and declining productivity growth. Experiments in the steel, auto, and electrical equipment industries are illustrative of both the origin and short-lived nature of these programs.

In the mid-1960s in steel, David MacDonald was defeated as president of the United Steelworkers by I.W. Abel, who ran on a platform of returning control of the union to the rank and file. Abel was swept into office by a rank and file demanding the right to strike over local issues and increased attention to shopfloor concerns (Betheil 1978). As a result, Abel endorsed the introduction of plant-level productivity and employment security committees in the early 1970s as a forum for the discussion of shopfloor issues of concern to labor and management. However, the committees were viewed as a threat by both plant management and local union representatives, and as a result became inoperative by the mid-1970s.

In autos, expressions of discontent by workers led to attempts by both employers and the UAW to address local shopfloor issues. The union had acknowledged workers' shopfloor concerns in a resolution passed at the 1967 collective-bargaining convention which read:

> The work place is not a penal colony; it must be stripped of its air of coercion and compulsion; imaginative new ways must be found to enable workers to participate democratically in decisions affecting the nature of their work. There is no reason why human and democratic innovation must continue to lag behind technological innovation in the plants and offices.
>
> (quoted in Bluestone and Bluestone 1992: 17)

QWL programs emerged in a number of GM and Ford plants in the early 1970s, but the majority of them faded away by the late 1970s, owing, in part, to the threat they posed to entrenched management and union bureaucracies. Plant-level industrial relations departments were so opposed to participation schemes at GM, for example, that initial responsibility for coordinating QWL programs was given over to the personnel department staff (Katz 1985: 76). The union's position on the effect of QWL committees on union structures was summarized succinctly in a 1980 statement by Irving Bluestone: "the provisions of the national agreement and of the local agreements and practices remain inviolable" (quoted in Katz 1985: 76).

In the early 1970s, realizing the shopfloor productivity problems facing its plants, General Electric also initiated a series of programs in an attempt to address the crisis of shopfloor contractualism.[2] First, the company introduced a program similar in spirit to a QWL scheme in which labor-management communication was enhanced through worker suggestions and joint meetings. Second, GE began experimenting in 1972 with work teams—an innovation that would become widespread in US manufacturing only much later in the 1980s. In its August 17, 1972, "Report For Managers," GE cited the possibility of "enriched jobs" and "greater participation in the operation of the line by employees" (6) as some of the effects of these experiments. Teams would offer workers "more freedom to pursue a course of action" and give workers "a sense of responsibility and enhance their self esteem" (7).

These early experiments were credited with reducing grievance rates and absenteeism, and with improving product quality and some shopfloor conditions (e.g., Katz et al. 1983). But with few exceptions the experiments were severely constrained in their scope and received little support from industrial relations departments and union officials—precisely those groups who were empowered during the rise of shopfloor contractualism, and whose positions would be threatened by a decentralization of shopfloor decision making.[3] Supervisors and foremen felt uncomfortable with the loss of authority that decentralized decision making might imply. Union officials were threatened by any challenge to the sanctity of the collective-bargaining agreement, and by the threat local agreements might pose for multi-employer bargaining structures. But the biggest immediate threat was to grievance committee representatives in unions and the staff in industrial relations departments, both of whose existence and identification rested on the older institutional arrangements.

Despite deep suspicion concerning management motives during a time of increased layoffs, rank-and-file workers, in contrast, were generally supportive of participation schemes of this sort. This is acknowledged even by the harshest critics of QWL programs (e.g., Parker 1985). Campaigns by management to convince workers of the mutuality of interest between employer and employee—which are often part of worker training classes prior to the introduction of participation schemes—may play a role in workers' willingness to experiment with these programs, as critics have maintained. But this explanation ignores workers' deep resentment of the system of shopfloor contractualism for its failure to allow sufficient worker participation in shopfloor decisions. Alternatives to this legalistic, contract-and-grieve approach to shopfloor governance are attractive to workers. (It is, nonetheless, also true that workers often fail to appreciate fully the challenge these alternatives pose for other revered features of postwar collective bargaining, such as the seniority principle.)

Through a slow process involving management and union administrative reshuffling and retraining, and a few key appointments to positions of leadership within management and union structures, the momentum for change was maintained in some firms. But case study evidence suggests that in most firms the early experiments with QWL programs had been abandoned by the late 1970s.

Participation schemes—often referred to this time as quality circles—reemerged in many industries following the recession of the early 1980s. The sustained erosion of domestic manufacturing in international markets, brought on in part by rising interest rates and the consequent high value of the dollar, provided the impetus for a more forceful attempt by firms to break the coalition of forces blocking structural change in the institutional arrangements of shopfloor governance. Rising international competition also caused the focus of reform to change; management searched increasingly for cost-cutting measures, while unions' interest turned more towards job security. Many of these new participation programs were explicitly designed as joint labor–management "cost study teams" (Klingel and Martin 1988). However, the thrust of the earlier failed reforms remained. One of the stated management goals was to decentralize shopfloor decision making in the

149

hopes of increasing worker participation and making better use of the knowledge of production possessed by workers.

A 1982 New York Stock Exchange survey of over 1,000 firms revealed that only 14 percent of firms reported possessing at least one human resource practice to stimulate productivity, and, among these, 44 percent reported having a QWL or QC program (Applebaum and Batt 1994: 62–3). (We have no systematic evidence on the extent of experimentation with joint-communication programs such as QWL prior to the early 1980s.) Evidence on the extent of experimentation with QCs during the mid-to-late 1980s is more plentiful, and the evidence suggests that QCs grew rapidly throughout the decade. Although not strictly comparable with the earlier survey, a 1987 survey found that 61 percent of responding firms possessed a QC (Lawler *et al.* 1989: 89). And a 1990 survey indicated that the number of firms with QCs stood at 66 percent (Lawler *et al.* 1992: 63).

However, there is also evidence to suggest that those companies reporting possession of a participation program rarely have a majority of their workers involved. The 1987 survey cited above found, for example, that in over 70 percent of firms reporting possession of a QWL or QC program, less than 20 percent of employees were active participants (Lawler *et al.* 1989: 90).

A study by Kochan, Katz, and Mower (1984) of QC programs in their early developmental stages sheds important light on the process of dispersion, the areas of resistance, the strength of the rank and file's desire for greater input into the nature of work, and the ultimate impact of the programs. Kochan *et al.* (1984) surveyed workers and local union officials in a small number of companies experimenting with QC programs.[4] They also conducted in-depth interviews with a small group of local union officials in a number of different industries. Their results confirm the existence of a rather slow pace of dispersion of participation programs throughout the plant workforce. Among the sample of rank-and-file workers, roughly half were participants in QC schemes even though the plants to which these workers were attached had been experimenting with these programs for an average of two to three years (Kochan *et al.* 1984: 98–105).

Local union officials cited management efforts to change work rules or practices, and the resentment and resistance of lower-level management staff to the participation schemes, as the primary reasons for the slow pace or actual blockage of progress in the spread of QC programs (Kochan *et al.* 1984: 146–8). This reveals both management's intention to utilize participation programs as a mechanism for changing the structure of industrial relations, as well as the ambivalence of supervisors and industrial relations staff to such changes. Interestingly, union officials who were interviewed generally stated that QC programs had a favorable effect on the rank and file's ability to communicate its concerns to both management and the union (Kochan *et al.* 1984: 134–8).

As a precursor to evaluating the impact of QC programs on aspects of the

job, the rank and file surveys asked workers about their interest in participating in decisions concerning the running of the plant. No less than 70
percent of the workers surveyed responded that they wanted "some say" over
"the way work is done," "the level of quality of work," or "how fast the work
should be done." No less than 63 percent said they wanted some say in the
choice of technology to be used on the job. These results offer very
compelling evidence of workers' interest in participating in the determination of shopfloor conditions.[5] By way of contrast, workers expressed little or
no interest in participating in decisions concerning such things as management salaries or promotions, and little interest in potentially important
issues such as plant closings and the investment of profits (Kochan *et al.*
1984: 106–12).

The results of the plant-level surveys of rank-and-file workers were
reported separately for each plant. Some plants had experimented with joint-
communication programs for only a short time, while others had more
lengthy experiences. Among the plants with roughly two years' experience
or less, the responses of participants were rarely significantly different from
those of nonparticipants on questions concerning the actual amount of say
workers had on the job (Kochan *et al.* 1984: 115–18). Thus, despite the
growth and popularity of QCs during the 1980s, participating workers
appear to have received little substantive increase in their ability to truly
influence shopfloor conditions by virtue of participation programs.

Nonetheless, in situations where the progress of participation schemes
was not successfully stalled by entrenched bureaucracies, cracks began to
appear in the structure of shopfloor contractualism. Katz's discussion of the
evolution of a participation program in an auto plant during the early 1980s
is illustrative in this regard. The committees initially tackled issues that did
not threaten existing contract language, such as lighting or the rearrange-
ment of work stations. But, before long, participation groups were engaged
in a feasibility study for the use of a robot in production and in discussions
concerning changes in the general layout of the shopfloor brought about by
alterations in product mix or the introduction of new technologies (Katz
1985: 80–5). Within the time span of several years, there was talk of instituting a far-reaching program of team production as a way of linking
participation groups to the formulation of workplace practice.

Team production

Comparisons between US and Japanese production methods in the 1970s
and 1980s revealed rather trivial differences in manufacturing technology,
leading to the conclusion that the production-cost advantages of our major
international competitor resided in organizational design (Abernathy *et al.*
1983). As both import competition and the performance of US firms in
foreign markets grew worse, domestic employers increasingly embarked on

shopfloor experiments of a more ambitious nature. What has emerged is an Americanized version of the Japanese system of shopfloor governance, known as "lean production" (Womack *et al.* 1990). Lean production incorporates QC discussion groups into a team production format, along with other attributes such as "just-in-time" production, total quality management (TQM), job rotation, and increased reliance on subcontractors for parts supply. Teams, however, are the heart of the shopfloor governance innovation of the lean-production model.

A number of alternative developmental paths towards team production emerged in US manufacturing during this period. In some industries, "greenfield" plants opened up with a new, cooperative organization of production. These were often nonunion plants within a largely unionized firm, as in rubber and autos. Team production methods were introduced as early as the mid-1970s in GM's nonunion plants in the south (Katz and Sabel 1985). In other cases, team production grew out of the earlier experiments with QWL programs and QC committees. In still other cases, where joint-communication programs had been successfully contained in their tendency to encroach on traditional company policy or contract language, teams emerged as last ditch efforts to prevent plant closures. Finally, Japanese firms have introduced teams in their manufacturing plants in the US

Team production in US manufacturing is largely an innovation of the 1980s. Less than 3 percent of a survey of firms in 1982 reported possessing production teams (Appelbaum and Batt 1994: 62–3). By 1990, survey results suggest that 12 percent of firms were experimenting with self-managed teams and an additional 4 percent planned to introduce them in the future (Appelbaum and Batt 1994: 65). As with joint-communication programs, reporting possession of a practice is not an indication that most, or even a majority, of a firm's workforce is involved in the practice. The same 1990 survey revealed, for example, that 59 percent of the firms reporting the existence of work teams had 10 percent or fewer of their employees involved in team production (Appelbaum and Batt 1994: 35).

A survey in 1992 by Osterman (1994) offers the best evidence to date on the extent of experimentation with workplace reorganization in the US. His findings suggest that over half of all manufacturing establishments employ work teams in at least one area of production, and that over 30 percent of establishments possess teams composing over half of the core production workers at their work sites. These findings suggest that the extent of experimentation with teams has risen significantly since the early 1980s, and that teams have become a central component in contemporary shopfloor experiments by employers.

Work teams can be found in a wide variety of manufacturing industries— steel, autos, rubber, electrical equipment, paper, foods, and chemicals. At the forefront of team production are such companies as General Electric, Procter and Gamble, Goodyear, General Motors, Corning, and Xerox (Parker and

Slaughter 1988; Appelbaum and Batt 1994). Unions have displayed a variety of approaches to the introduction of team production in unionized plants, from outright endorsement to resistance at all costs (the latter is especially true of the skilled trades unions). After initial resistance, the UAW, for example, explicitly endorsed teams in national contracts with Ford and GM in 1987. The United Electrical Workers and the Oil, Chemical, and Atomic Workers, on the other hand, have expressed strong opposition to team production.

In manufacturing, the auto industry's experience with work teams is the most well known. Parker and Slaughter note that by March of 1988 work teams were in place in at least "17 General Motors assembly plants, in six Chrysler plants, in Ford's Rouge Steel operation and Romeo engine plant, and in all of the wholly or partially Japanese-owned plants (Nissan, Honda, Mazda, Diamond-Star and NUMMI)" (Parker and Slaughter 1988: 4). Teams are also prevalent in components plants in the auto industry and are a critical element in the production process at GM's Saturn plant. Team production has emerged in autos via a number of different paths, from their introduction in the mid-1970s in nonunion plants to the cooperative effort between union and management in the Saturn project.

The diversity of experience with team production makes it difficult to describe in general terms. Most team production systems, however, contain the following attributes: a dramatic reduction in job classifications, sometimes to a single classification for all production workers; workers becoming skilled in a much fuller range of production activities; work teams composed of between ten and fifteen workers who meet weekly to discuss such issues as efficiency, quality of product, and the job assignments of team members; a team leader (typically a union member in union plants) from the ranks of workers and a group leader (a salaried, nonunion employee) from the ranks of management as coordinators of team activity; and a "pay-for-knowledge system" which encourages workers to acquire different skills in the plant by awarding increased wages for skill acquisition. Workers and supervisors are encouraged by this structure to resolve shopfloor disputes speedily, with a minimum of bureaucracy.

Team production has been generally well received by workers. For those who have labored under the system of shopfloor contractualism, the absence of bureaucratic procedures governing shopfloor production and dispute resolution can indeed be inviting. Levine reports, for example, that an overwhelming majority of workers at the NUMMI plant in Fremont, California—many of whom worked in the plant under GM management before the joint venture between GM and Toyota began in 1984—consistently report being "satisfied with (their) job and environment" (Levine 1995: 13) under the team production system in place in the plant.

Similar sentiments are offered by workers at the GM–Suzuki joint venture (CAMI Automotive) in Ontario, Canada (Robertson et al. [n.d.]:

26). A survey in 1990 revealed that a majority of workers expressed the view that teams "helped them to do their jobs better" (66 percent); "gave them a say about their jobs" (73 percent); and "allowed team members to act together to express complaints" (92 percent) (Robertson *et al.* [n.d.]: 21). These favorable views of work teams appear to translate into active participation in the affairs of teams as well. Workers at CAMI regularly participate in offering recommendations on how to improve productive efficiency; survey evidence reveals that over 70 percent of the workforce submitted suggestions for improving productivity during the plant's second year of operation (Robertson *et al.* [n.d.]: 21).

However, workers' positive responses to working in teams mask a deep distrust of management's motives in introducing teams, as well as concern about the impact of team production on the quality of certain shopfloor conditions. CAMI workers, for example, while generally pleased with certain aspects of team production, nonetheless report by an overwhelming majority that "teams work more for the good of the company than workers," and by a near majority that teams are "a way to get people pressuring one another" in production and "a way to get people to work harder" (Robertson *et al.* [n.d.]: 27). At NUMMI a People's Caucus within the union won the local presidency in 1991, in part because of its critical stance on the issue of team production (Levine 1995: 19).

Other survey results seem to confirm this evidence on workers' views of teams. Kochan *et al.* (1984) surveyed a group of workers from one plant where team production had emerged out of earlier experiments with a joint-communication program. At the time of the survey, workers had only a very limited experience with work teams since the experimental programs had been in operation for less than a year. Teams had advanced to cover only a portion of the plant workforce, thus allowing a comparison of responses by team participants and nonparticipants. On eight of eleven questions concerning how workers viewed their work (including issues such as whether the job is "meaningful," "requires that workers learn new things," and "gives the worker a sense of his or her impact on the final product or service"), participants' responses were significantly more favorable than those of nonparticipants (Kochan *et al.* 1984: 116). However, when asked how much actual influence workers felt they had over various aspects of their work (e.g., "the way the work is done", "the level of quality of the work", or "the use of new technology on your job"), in only one of seventeen areas of concern ("who should do what job in your group or section") were participants' responses significantly different from those of nonparticipants (Kochan *et al.* 1984: 113).

Moreover, more negative views of teams by workers are not difficult to come by in the literature. Parker and Slaughter's case study of GM's Factory 81 plant, where torque converters are built as part of the giant Buick complex in Flint, Michigan, is illustrative. What was touted as an ambitious experiment in worker autonomy through teams—one of the first such experiments at GM with a group of union workers working under a traditional bargaining arrangement—gradually turned into something quite different. The promise by management was a plant with no shop rules, no time clocks, and where workers could set production standards. Although jobs were designed before work began in the new plant, workers were initially allowed great freedom in the scheduling of work and in job assignments. But before long even these minimal freedoms had been eroded by management (Parker and Slaughter 1988: 192–5). A local union official who had been an early supporter of the move to cooperative relations with management summarized the experiment in this way:

> The bottom line . . . is this: anything joint should be 50–50. But in reality it's 51–49. When it's a question of "quality, cost, schedule," *they'll* make the bottom line decision. It's their plant.
> (Parker and Slaughter 1988: 195)

Workers at the GM-Van Nuys plant led one of the earliest and most well-publicized campaigns to convince GM to be more forthcoming about the precise nature of team production when GM was proposing it as a solution to the plant's productivity problems in the mid-1980s (Mann 1987). After work teams were eventually adopted at the plant, workers responded with a variety of attempts to enhance their shopfloor power within the team structure (Parker and Slaughter 1988). Continued friction over teams was arguably responsible for the subsequent closing of GM-Van Nuys.

The Van Nuys' workers' reaction to their experiences with team production is not an isolated incident. Cappelli and McKersie (1987: 455) report, for example, that strikes at two of GM's midwestern assembly plants have been associated with favoritism in job assignments and pay-for-knowledge increases. Katz (1985: 91) cites the example of GM's plant in Oklahoma City, which until 1979 was nonunion and one of GM's southern strategy plants. During the negotiation of the first local contract, the workers pushed for the elimination of team production. Ultimately, what emerged in this plant was the negotiation of job classifications that are somewhat fewer in number, but nonetheless similar in other respects to the traditional collective-bargaining structure.

Other kinds of evidence coming out of the experiments with team production are equally difficult to assess. Take, for example, the evidence on grievance rates. One of the most troublesome features of the system of shopfloor contractualism was the backlog of grievances concerning shopfloor disputes. Plants using team production typically decentralize shopfloor dispute resolution rather dramatically. For example, while the NUMMI–UAW contract contains a number of steps for resolving workplace disputes, much like a traditional one, the contract also clearly states that the first step for resolution is the work team (or Group), and goes on to state that "The Company and the Union shall encourage all employees to attempt to resolve problems [the term used for grievances in the contract language] within the Group using problem-solving methods" (Parker and Slaughter 1988: 115).[6]

By all accounts, the attempt to address shopfloor concerns promptly has been enormously successful. It was not unusual for the GM-Fremont plant (before NUMMI) to have a load of over 5,000 grievances, and a backlog of unresolved grievances of over a 1,000 (Brown and Reich 1989: 28). During the three years of NUMMI management, formal grievances have fallen dramatically. For example, over this period only four grievances were filed for arbitration (Brown and Reich 1989: 29). When workers were asked by Parker and Slaughter how the "problem-solving" procedure at NUMMI compared to the grievance procedure at GM-Fremont, not one wished to return to the traditional system. Workers stated that:

> At GM . . . once the grievance was written, you never heard again . . . the grievance took forever . . . it just got kicked upstairs . . . general supervisors would always back the foremen . . . you never won . . . if you won it was too late to do any good.
>
> (Parker and Slaughter 1988: 109)

Not all that glitters is necessarily gold, however. Grievances concerning job standards in traditional contracts in the auto industry were not covered by arbitration and were therefore strikeable issues. This is not the case at NUMMI. Moreover, NUMMI management has removed the possibility that the decentralized team structure—which bears a certain resemblance to shopfloor governance under fractional bargaining—would allow for precedent-setting shopfloor changes by workers through the grievance procedure. The company has adopted measures similar to those employers utilized to attack the earlier system of fractional bargaining—for example, maintaining in the contract a clause that any resolution at the first stage of the problem-solving process "shall not set a precedent or a binding past practice on either party" (Parker and Slaughter 1988: 115).

Finally, it should be acknowledged that reduced grievance rates are not a sure sign of increased shopfloor contentment. Low grievance rates may reflect a tighter control on expressed discontent by management and a loss of worker voice. In other auto plants the number of filed grievances has also fallen under team production, but workers there blame the reduction on the unresponsiveness of committeemen under team settings (Parker and Slaughter 1988: 132–3).

Evidence on rates of worker absenteeism carry similar mixed messages. Absenteeism rose dramatically during the crisis of shopfloor contractualism in the late 1960s. It was not uncommon for the rate of worker absences to reach 20 percent of the workforce at GM-Fremont before NUMMI was formed. In 1992, by contrast, NUMMI had a worker absence rate of 3 to 4

156

percent (Levine 1995: 13). Does this reflect increased worker satisfaction? It is hard to tell. Team production grants final responsibility to the team for seeing to it that production goes smoothly; a worker absence can therefore mean that other team members will have to cover the absent worker's job. Thus, team production focusses peer pressure on each worker not to miss work (Parker and Slaughter 1988).

Indeed, despite the satisfaction with work teams occasionally voiced by workers, there is very little evidence to suggest that team production truly empowers workers in production. This lack of genuine empowerment is revealed in a number of ways. While job design and work standards are not subject to contractual regulation under team production systems, neither are they open to substantive rank-and-file control. Job design and work standards are determined largely by management. Workers are then sometimes given the freedom to decide how job tasks are combined to form a job. But more often workers are simply given the ability to choose the assignment of team members to the various jobs that compose the work of the team.

The process by which jobs are designed and work standards are determined under team production typically runs as follows:[7] Industrial engineers are responsible for breaking the process of production into a series of "transferable work components," each composing the smallest combination of acts for which it would be impractical to have more than one person responsible in the course of production. Engineers also recommend the most efficient procedures for carrying out each "transferable work component" as well as the time standards for each, to a tenth of a second in some cases (Graham 1995: 79). Summing up the time standards for each "transferable work component" thus produces the total time allotted for producing one unit of the product.

Given the number of workers in a team and the team's position in the process of production, the number of work components for which each team is responsible is then determined. Team leaders enter the process at this point. Using manuals which give recommended time standards for each of the team's assigned work components, team leaders design the various jobs of the team by combining work components in the most efficient manner possible. In this way the tasks attached to the various jobs of a given team are determined.

The basic layout, tools and equipment, and the specific steps and sequences of worker moves is strictly spelled out. This is so that standards can be changed through "kaizen" activities—that is, "eliminating the unnecessary" (see Robertson et al. [n.d.]: 7; Graham 1995: 75–9). Work teams are asked to adhere to these standards and then encouraged to offer suggestions on how the standards might be altered to increase productive efficiency. Adjustments can be made to the combination of job tasks for any given job, to the assignment of work components to any given team, and even to time standards should the operation in practice not run as smoothly as, or run even smoother than, seemed possible on paper.

The opportunity for workers to participate in adjustments of this sort produces the possibility for continuous improvements in productive performance, but it also creates a setting in which there is competition both between workers within a team and across teams in the plant. This is a second way in which team production limits workers' shopfloor power. To the extent teams can foist off work components onto other teams, their work lives will be made easier. The same holds true for individual workers within teams. The incentive for workers to compete with their peers exists not only in the strategic adjustment process during which work components are assigned and reassigned to workers and teams, but also through the collective responsibility teams are given for completing their assigned tasks. This collective responsibility produces incentives for teams closely to monitor individual team members' absences (mentioned above) and work intensity while on the job.

Teams are the informal work groups of lean-production plants, and competitive and mutual monitoring incentives reduce the power of informal work groups, both individually and collectively, thereby reducing workers' ability to cooperate in an effort to garner and protect shopfloor improvements. This is perhaps why its critics view team production as an ingenious new method of management control that reduces worker resistance to management goals through peer pressure and reduced solidarity (Dohse *et al.* 1985; Parker and Slaughter 1988).

The power of workers to influence shopfloor conditions is also limited by the divided loyalties of team leaders. Team leaders are workers in that they are required to know all of the jobs in the work team and to fill in for absent team members. In unionized plants, they are generally members of the union, and are the equivalent of shop stewards or committeemen under more traditional arrangements for dispute resolution. But they are also management in that they supervise workers' performance and have some say in the design and assignment of workers to jobs. The decentralization of managerial authority characteristic of team production means that team leaders have a fair amount of discretionary power over workers.

Many workers appear to feel that the divided loyalties of the team leader, and especially the increased discretionary power to make managerial decisions, make it difficult for the person holding the position to act as a proper shop steward. Moreover, many resent the fact that team leaders are not democratically chosen by team members. Team leaders are typically chosen by management, or jointly by union and management in organized plants. In many team production plants, movements have surfaced among workers to have team leaders chosen by the workers, just as shop stewards are chosen under traditional arrangements. At CAMI, for example, while workers are generally supportive of the need for a team leader, they express the desire by an overwhelming majority (over 75 percent) for the team leader to be elected by team members (Robertson *et al.* [n.d.]: 30).

A further indication that team production limits the shopfloor power of workers is the association teams have come to have with increased worker stress. Despite the claimed autonomy of work teams, very little of this stress appears to be self-imposed. "Kaizen" activities that are performed by workers on their own initiative, and which result in a faster work pace and thus more stress, may be viewed as self-imposed. But team production plants "kaizen" in many ways that are not under the control of workers, from periodic speedups to expose the weakest links in the process of production (Graham 1995: 105) to roving teams of management who "kaizen" independently of work teams (Robertson *et al.* [n.d.]: 22).

Stress is also produced by the goal set out by management in many lean-production plants to achieve zero defects. In autos this desire is referred to as producing 100 percent "direct runners"—that is, having every car prepared to pass final inspection as it rolls off the line. Workers are thus pushed to make repairs on the line. The use of "just-in-time" production, in which stockpiles of parts and supplies are kept to a minimum to conserve on inventory holding costs, also produces stress in that even if the worker is not personally responsible, an inability to complete job tasks leads to the line being halted and to the worker's team being held responsible for the stoppage.

Team production is a system that builds on workers' natural desires to produce a quality product in surroundings that are reasonably pleasant and under conditions of relative autonomy. While the team concept acknowledges these desires, it rarely truly fulfills them. It is a system of management control in which the *responsibility* for producing is squarely placed on workers' shoulders while the production goals and job standards are dictated by management.[8] Experience to date suggests that in exchange for an increase in worker stress, an ideological climate that promotes competition between workers (within and across work groups, as well as across plants and firms), and fewer contractual shopfloor protections, workers receive a quicker turn-around time for the resolution of workplace disputes, and a limited say in the nature of work.

PRODUCTIVITY AND SAFETY UNDER LEAN PRODUCTION

Statistics on the recent trajectory of labor productivity growth in manufacturing lend suggestive evidence in support of the claim that experiments in the US in workplace reorganization, which began with great force in the 1980s, have produced positive results. Manufacturing output per worker-hour grew at a paltry 1.7 percent per annum during the 1970s, but rebounded to an annual average of 3 percent during the period 1981–92. This exceeded labor productivity growth in Germany and was equivalent to that for Japan during the same period (*Economist* 1993: 67).

However, even better evidence exists on the productivity impact of the recent workplace reorganization. Given the emphasis placed on improving labor productivity in these employer experiments, it is not suprizing perhaps that there exist a number of empirical studies exploring the relationship between joint-communication committees, team production, and shopfloor productivity. In what is arguably the best survey of this literature, as well as the much larger literature on the relationship between worker participation and productivity, Levine and Tyson conclude:

> [Employee] participation usually has a positive, often small, effect on productivity, sometimes a zero or statistically insignificant effect, and almost never a negative effect. The size and significance of the effect are contingent on the type of participation involved and on other aspects of the firm's industrial relations environment. Participation is more likely to have a positive long-term effect on productivity when it involves decisions related to shopfloor daily life, when it involves substantive decisionmaking rights rather than purely consultative arrangements (for example, quality circles), and when it occurs in an environment characterized by a high degree of employee commitment and employee–management trust.

> (Levine and Tyson 1990: 183)

In perhaps the most careful study to date on the effect of participation on productivity in American manufacturing, Ichniowski *et al.* (1994) find that steel plants utilizing worker participation through joint-communication committees and teams achieve marginally higher levels of labor productivity than steel plants without employee participation. The authors find, however, that when the package of reforms includes such things as profit sharing and employment security, in addition to worker participation, the productivity boost is substantial.

The empirical evidence thus suggests that efforts to introduce the lean-production model into American manufacturing have had a marginally positive impact on productivity, which is boosted significantly in the unfortunately all too seldom cases where employers also grant workers employment security and substantive rights of participation in production. Interestingly, an issue not taken up by any of the existing studies is the precise explanation for the productivity boost from experiments with lean production. Is it greater cooperation between labor and management in the elimination of productive inefficiencies, the benefits of which redound to both groups mutually? Or is it instead the result of cutting corners on workers' shopfloor conditions, resulting in speedups and increased stress?

There is certainly ample evidence to suggest that the intensity of labor effort increases under team production and, therefore, that at least some of the productivity boost is the result of speedups. For example, while NUMMI witnessed a roughly 50 percent increase in productivity over that of its former operation as GM-Fremont (Brown and Reich 1989: 32), if the work pace under its former incarnation was anything comparable to that of other GM facilities, part of the explanation for the higher productivity at NUMMI is clearly increased labor intensity. Treece (1989: 80) found in a comparison of NUMMI and GM's Linden plant, for example, that workers work fifty-five seconds out of every minute at NUMMI, but only forty-five seconds out of sixty at GM-Linden.

Similar suggestive evidence emerges from case studies of CAMI and various Japanese auto transplants in the US. At CAMI, for example, one third of interviewed workers felt the pace of work was too fast (Robertson *et al.* [n.d.]: 31). Babson's survey of workers at the Mazda plant in Flat Rock, Michigan, reveals that three-quarters of the workforce felt that their work pace was so intense that they would be either injured or worn out before they reached retirement (cited in Graham 1995: 12). Most Japanese auto transplants in the US attempt to run as close to 60 seconds per minute of work as possible (Fucini and Fucini 1990: 37), allowing them to assemble vehicles with an average of 21.2 hours of labor compared to 25.1 hours for comparable US plants (Womack *et al.* 1990: 92).

Parker and Slaughter offer evidence of a different sort on the work pace under lean production—they simply describe a typical production job at the NUMMI plant:

> In 59 seconds, inspector Richard Aguilar has to get in and out of the car and check to make sure each contains the items specified on a form for that particular car. Each item is also checked to see that it operates correctly. The list includes: headlights; high beam; turn signals; back lights; side marker lights; parking lights; radio; speakers; heater; air conditioner; dome lights; air ducts; steering wheel; console; dash; shift lever; check upholstery for color, cleanliness, tightness, damage; check headliner for tightness and damage; check garnishes (moldings which cover joints).
>
> (Parker and Slaughter 1988: 105)

What about the relationship between lean production and workplace injuries? The manufacturing injury rate has continued its rise over the past few decades, from roughly 4.0 injuries per 100 workers in 1972 to 5.3 in 1990. However, the role of workplace reorganization in the increased injury rates of this period must compete with several alternative explanations (Smith 1992). During these years, the Occupational Safety and Health Administration (OSHA) increased the fines it assesses on firms that keep inaccurate injury records, leading firms that once made a practice of disguising their true injury rates to report more accurate, and hence greater, injuries. There has also been a significant rise in benefits under most state workers' compensation systems during this period, something that has been found to increase reports of injuries by workers.

However, there is also evidence to suggest that some of the rise in injury rates may be attributable to employers' efforts at workplace reform. To the extent speedups are indeed associated with the introduction of lean-production systems, the result may well be increased injuries. A case for there being a positive relationship between speed of production and shopfloor injuries has been recently made most persuasively in a series of articles in the *Wall Street Journal*, the central conclusion of which is that increased job-related injuries and illnesses have much to do with people being "pushed to produce." At the John Morrell & Co.'s Sioux Falls, South Dakota, meat-packing plant the speed of the line has been increased in some departments by as much as 84 percent in the 1980s, and plant-wide injuries increased over this period by 51 percent (*Wall Street Journal* 1989: 41). Soaring injury rates among poultry workers can also be related to the near doubling of the line speed—from the high fifties to ninety chickens per minute—over the past fifteen years (*Wall Street Journal* 1994: A10).

Lean production methods have other attributes, besides increased work intensity, that may lead to increased injuries: a greater use of temporary workers who may be less familiar with production; a reduced number of maintenance workers who are responsible for safety lock-down during equipment failure; and increased subcontracting of parts suppliers which results in ill-fitting parts that require greater tugging and twisting to install. Finally, and perhaps most importantly, workers in lean-production plants are made to feel that they, and not the company, are responsible for safety.

Case study evidence from lean-production plants in the US sheds some empirical light on the relationship between workplace reform and shopfloor accidents. Following the 1993 model change at NUMMI, there was a 12 percent increase in worker absences due to health and safety problems. The apparent cause was the prescribed time standards for job tasks under the new model. Although injuries began to increase long before the new standards were met, the company failed to respond to the problem and continued with its plan to increase line speeds in order to meet prescribed standards. The California Occupational Safety and Health Administration was finally summoned to the plant, resulting in a citation concluding that "serious employee injuries due to repetitive stress, as well as employee symptoms of impending stress injury increased alarmingly" following the model changeover (quoted in Levine 1995: 33).

Similar stories emerge from the Japanese transplants in the US. At Mazda, injuries increased by 50 percent between June and September of 1988 as the plant approached full production (Fucini and Fucini 1990: 175). Berggren *et al.* (1991) visited a number of Japanese transplants in the US. and found growing health and safety complaints related to the intense pace, repetitive job tasks, and long hours. At Mazda they found extremely high levels of repetitive strain injuries and an injury rate three times the level of other US auto plants (55).

Wokutch's (1992) recent research offers yet another glimpse into the level of workplace safety in a representative Japanese auto transplant. The plant is an assembly operation, utilizing the various features of Japanese shopfloor governance such as teams, a just-in-time inventory system, TQM, and quality circles. Wokutch cites anecdotal evidence suggesting that productivity at this plant is significantly higher than US-owned assembly plants (145). Although workers and line management are Americans, and workers are organized by the UAW, the plant's basic approach to health and safety is modeled on that of its plants in Japan. After a couple of years in operation, the injury and illness performance of this plant compared very unfavorably to comparable plants in the industry.

For 1988, Wokutch found that the injury and illness frequency (44.4 per 200,000 hours worked) was 91 percent higher than the rate for the industry, and 66 percent higher than the rate for similar auto plants employing at least 1,000 workers (Wokutch 1992: 192). Strains, sprains, and cumulative trauma disorders (CTDs)—which can be attributed to the stress of the production system—accounted for a large share (almost 50 percent) of the reported injuries and illnesses in the plant for 1988 (195). For CTDs alone, the rate was roughly five times as high as the rate for comparably large auto plants in the industry (195).

To what extent does the evidence in this case study hold true for the broader class of establishments experimenting with lean-production techniques? Data exist which make it possible to offer suggestive evidence on the relationship between lean production and workplace health and safety across US manufacturing establishments. The Organization of Work in American Business Survey is a survey of establishments, conducted in 1992, to measure the extent of workplace reform efforts in the US.[9] Establishments were asked, among other things, about the degree to which workers who compose the core occupational group in their establishments were organized into teams, utilized quality circles, employed TQM techniques, engaged in job rotation, etc. Extracted from the survey for analysis were those manufacturing establishments whose core work groups were blue-collar occupations. Using each establishment's reported SIC industry code, the average annual change in injury rates between 1986 and 1991, for the industry in which each establishment resides, was calculated and imported into the data set.[10]

The average injury rate increase for this sample of establishments was 4 percent, a trend that is consistent with that of manufacturing as a whole (Smith 1992: 560). The results of a simple regression of injury rate changes on various aspects of the lean-production model of workplace reform are contained in row 1 of Table 7.1. Additional control variables include the number of workers in the establishment, the percentage of the plant workforce composed of female workers, the number of years the establishment had been experimenting with the lean-production approach (as measured by experience with quality circles), the percent of the workforce in blue-collar jobs, and the percent of the blue-collar workforce covered by a collective-bargaining agreement.[11]

163

Table 7.1 Suggestive evidence on the relationship between lean production and workplace health and safety trends in manufacturing (sign and significance reported)

Regression analysis

Dependent variables	Independent variables					N
	QC	Teams	TQM	Rotation	Skill Change	
(1) Injuries$_{1986-91}$	+	−	+*	+		295
(2) Injuries$_{1986-91}$	+	−	+*	+	+**	295
(3) CTDs$_{1986-91}$	+*	−***	−	+		295
(4) CTDs$_{1986-91}$	+*	−***	−	+	+	295

*   significant at the .10 level (one-tailed test)
**   significant at the .05 level (one-tailed test)
*** significant at the .01 level (one tailed test)

The results in row 1 suggest that lean production has contributed significantly to the worsening of workplace safety in manufacturing over this period. While three of the four lean-production variables—namely, quality circles, teams, and job rotation—are not statistically associated with injury rate movements during these years, the extent of experimentation with TQM is significantly related to rising injury rates.[12] Roughly 10 percent of the average rise in injury rates over this period can be accounted for by the spread of TQM.[13]

It is not surprising that, of all the aspects of the lean-production model, TQM is the one significantly related to increased workplace injuries. Whereas quality circles and teams are, at least in part, directed at increasing worker participation, TQM focuses almost entirely on process changes such as the elimination of "wasted" time and motion, increased throughput (i.e., speed), simplification of tasks, and reduced cycle times. It is a top-down management approach to reengineering the production process reminiscent of the Taylorist principles of scientific management which were introduced during the early twentieth century (Appelbaum and Batt 1994: 88–91).

Another important feature of contemporary workplaces that may influence injuries is rapidly changing skills due to the changing technology and organization of production.[14] The development of new skills on the job can lead to increased workplace injuries in the short run, as workers adjust to their new job tasks. Once greater confidence and familiarity are attained, however, the impact on injuries ultimately rests on the degree of danger inherent in the nature of the new technologies and organizational arrangements. To the extent lean-production establishments are subject to more rapidly changing skill demands, the failure to control for changing skills in the workplace may bias the results presented in row 1.

In row 2, a variable capturing the degree to which the skills of workers in the core work group have changed significantly in recent years is added to the regression equation. The introduction of this variable does not alter the findings presented in row 1. (A separate analysis reveals that there is surprisingly little correlation between changing skills and lean-production workplace reforms.) The positive and highly statistically significant relationship between changing skills and injuries is nonetheless striking on its own terms. Changing skills in the workplace account for, on average, roughly 20 percent of the average annual rise in workplace injury rates over this period.[15]

Injury rate data do not capture the strains and other health-related disorders increasingly associated with jobs that possess short cycle times and which require repetitive movements and a fast work pace. These "repetitive motion disorders" or "cumulative trauma disorders" (CTDs) as they are variously labeled appear as a separate entry in government statistics on occupational health and safety. The rate of CTDs has grown dramatically in recent years, quadrupling in manufacturing, for example, between 1986 and 1991.[16] While CTDs are fast becoming a significant health concern, they affect a much smaller percentage of the manufacturing labor force than do injuries. For the period 1986–91, the average yearly lost-workday injury rate was roughly 50 workers out of 1,000; the average CTD rate was 6 out of 1,000, or about one-tenth the injury rate.

Part of the recent growth in reported CTDs is arguably a matter of increased awareness and acceptance of such disorders as legitimate workplace-related illnesses. Thus, growth rates in reported CTDs across industries may reflect differences in the accepted legitimacy of such disorders—owing perhaps to differences in the power of workers to influence the recording of injuries and illnesses or to differences in the progressivity of management thinking on the issue—rather than reflecting an accurate account of across-industry experience with CTDs. Much of the case study evidence suggests, however, that CTDs are indeed growing and that they are a particularly important problem in lean-production settings. Thus, while keeping in mind the possible mismeasurement issues, it seems important to explore the relationship between lean production and CTDs.

In rows 3 and 4 of Table 7.1 we replicate the earlier analysis of injury rates for the case of cumulative trauma disorders. For our sample of establishments, the annual average rate of change in CTDs over the period

1986–91 is 65 percent, which is consistent with the very rapid increase in reported CTD rates in manufacturing more generally. The results of row 3 reveal that quality circles are positively associated with the growth of CTDs and that teams are negatively associated with CTD growth. The results of row 4 reveal that the introduction of a variable capturing recent changes in worker skills due to technological and organizational change does not significantly alter the findings presented in row 3. Indeed, skill changes themselves appear to have little bearing on the rate of growth of CTDs across industrial establishments.

The negative association between team production and CTD growth runs counter to what one might expect given some of the existing case study evidence. However, the result is quite plausible. The greatest freedom that work teams are granted is the self-determination of job assignments. To the extent CTDs are more likely to occur where workers engage in lengthy stays at very repetitive job tasks, teams can reduce the incidence of CTDs by rotating workers through the various job tasks performed by the work team. (The job rotation variable, which represents management's assessment of the extent of worker job rotation, will fail to capture job rotation that is initiated by teams unbeknownst to management.)

The positive association between quality circles and CTDs is more difficult to explain. Quality circles may provide a forum in which management can mollify worker complaints about increased pace and shorter cycle times for jobs—productivity enhancing alterations that may also increase the incidence of CTDs. Alternatively, the positive association may be related to the greater likelihood of an accurate reporting of CTDs in settings where workers are given the opportunity to discuss both the quality of the product and the quality of worklife.[17] Clearly, more research is required before we will fully understand the underlying explanation for this finding.

What is the overall impact of lean production on the rate of growth in CTDs? Appendix Table 7.1 gives the estimated coefficients and sample means for the empirical results presented in row 3. These reveal that quality circles account for roughly 8 percent of the average annual rise in CTD rates over this period, and that teams reduced the growth rate in CTDs by roughly 11 percent. Looking only at these two aspects of the lean-production model—the two that appear to be significantly associated with CTDs—the effect of the experiments with lean-production methods has been to reduce the growth of CTDs on the order of about 3 percent.

What has been the impact of lean production on the trajectory of overall workplace health and safety during the years 1986 to 1991? Given the greater prominence of injuries in the overall numbers of safety and health problems in the workplace, when the effect of lean production on injury rates is added to its effect on CTDs, lean production is found to have contributed positively, to the worsening health and safety record of American manufacturing in the recent period.

CONCLUSION

The past few decades have witnessed increased experimentation with the lean-production model in US manufacturing. These experiments have been initiated by employers, and are, in part, a reaction to the crisis-ridden aspects of the system of shopfloor contractualism which emerged in many manufacturing firms during the 1960s. The lean-production model strives to recapture certain virtues of the decentralized system of fractional bargaining of the 1940s and 1950s, while at the same time constraining workers' power to influence shopfloor conditions through informal work groups and shop stewards.

Empirical evidence suggests that lean production has contributed modestly to both increased labor productivity and worsened workplace health and safety. Part of the boost in productivity is ostensibly due to the increased intensity of labor effort, which in turn appears to be a likely contributor to worsening workplace safety levels. Thus, the American version of lean production is not an unequivocal improvement on the system of shopfloor contractualism. Despite the improved labor productivity, the costs to workers in the form of rising injury rates, increased worker stress, and a heightened work pace should lead us to wonder whether there exists a superior alternative to lean production as a set of institutional arrangements for shopfloor governance.

8

A VISIT TO SATURN

The Saturn Corporation was announced by General Motors in January of 1985 as a project to produce a compact vehicle that would compete with foreign producers and be a world leader in quality, cost, and customer satisfaction. Saturn was also meant to be the breeding ground for a new set of industrial relations principles for ailing GM; after being ironed out at Saturn, these principles would spread rapidly to the remaining GM production facilities. A revolutionary agreement with the UAW, formal independence from the larger GM management structure, and years of careful planning went into the project before the first car rolled off the line in July of 1990, in the town of Spring Hill, Tennessee.

The structure and practice of industrial relations at Saturn are pathbreaking for a US corporation. To begin with, many of the supporting features of the lean-production approach which are present in Japan, but absent from many of the employer experiments domestically, are also present at Saturn. Five percent of work time is devoted to worker training; 20 percent of the compensation package is based on success in exceeding plant-level productivity and quality goals; workers are guaranteed life-time employment; and the formal arrangements of shopfloor contractualism—lengthy bargaining agreements and the adjudication of disputes through a legalistic grievance procedure—are eschewed in favor of joint labor–management decision making and ongoing problem solving.

Off-line problem-solving circles and on-line work teams of approximately fifteen workers each are features that Saturn shares in common with other domestic manufacturers experimenting with workplace reorganization. At Saturn, however, self-directed teams were given substantial autonomy in decision making, both in the planning stages of production and in the first few years of operation. Teams took responsibility for quality control, ordering and handling material, controlling their own budgets, and even hiring new team members. The original goal, as put by a company vice-president, was to have "employees accept ownership for the direct labor functions they perform" (quoted in Solomon 1991: 72).

A truly unique feature of the governance structure at Saturn is the "partnering" arrangement between management personnel and union leaders at both the staff and operating levels (Rubinstein *et al.* 1993: 344–5). For example, the vice-president of people systems at Saturn, like all other occupants of senior staff positions, is partnered with someone from the UAW who is meant to share jointly in all decisions associated with the office. Foremen and superintendent positions, suitably altered to reflect the greater autonomy of teams at Saturn, are filled by two employees of the company—one from management and one from the union. On the shopfloor, teams are organized into larger modules of roughly 100 production workers, and a pair of advisors is assigned to give guidance and resources to teams within their module. The "worker representatives" in this pairing process are not elected by the rank and file, but rather appointed by the union and management.

When the Saturn project was announced in the mid-1980s, the partnering arrangement was hailed by UAW President Owen Beiber as something that would make the union a "full partner in all decisions from the shop to top management," thereby granting workers "a degree of co-determination never before reached in US collective bargaining" (quoted in Sherman 1994: 86). A closer look at the structure and practice, however, reveals that the distribution of decision-making power between labor and management at Saturn rests less on a formal set of worker rights, reminiscent of co-determination in some European settings, and more on the informal norms and "ethos" of the institution.

In order to ensure that the informal norms would produce labor–management cooperation and that the ethos would foster joint decision making, GM committed enormous time, energy, and funds to the recruitment of personnel. The recruitment process for production workers alone was arguably like no other in the history of manufacturing. In the winter of 1988 a recruiting department launched its hiring campaign, visiting the first of 136 targeted GM plants in search of worthy candidates. Follow-up phone interviews were conducted with interested and effective candidates, and, if that hurdle was jumped, prospective workers were given a plane ticket to Spring Hill where more interviews and a series of tests were conducted. By one estimate, perhaps one-third of those interviewed by phone were invited to visit the plant, and perhaps one-fifth of those who visited were finally offered a job (Sherman 1994: 199).

An equally important contributor to the behavioral norms of labor and management at Saturn was the training process. Production workers, for example, underwent from 300 to 350 hours of training, involving not only the attainment of competence in a variety of tasks within their designated team operation, but also team dynamics, consensus decision making, trust, respect, and the fundamental relationship between responsibility and accountability ("you have to be responsible to be accountable" was the message). One course, utilizing methods of self-esteem building and team cooperation, required workers to both individually and collectively overcome fears in accomplishing a set of physically and emotionally demanding tasks (Sherman 1994: 210–11).

The luster has faded a bit from some of the project's original intentions. The early practice of maintaining autonomous, "self-directed" teams has changed, as teams have been forced to accept greater direction from above (Sherman 1994: 271). Recruiting is now not nearly so selective, in part because the UAW has demanded that new recruits come from the ranks of laid-off GM workers. And while training is still extensive, there is currently less devotion to behavioral transformation through self-esteem building and the dynamics of team cooperation (Sherman 1994). Nonetheless, the Saturn project arguably remains the most advanced experiment in joint labor–management decision making and cooperation that exists in the US today.

It is for this reason that I decided to visit the plant, in hopes of seeing, to the extent possible in a short visit, how the experiment was going from the point of view of production workers. After sending an initial letter of intent and then receiving a follow-up interview with Jennifer Graham in public relations, I was invited to visit Saturn on the afternoon of September 7, 1994, during which time I was to be given a plant tour and then allowed to roam the shopfloor, accompanied only by Ms. Graham, to conduct informal interviews with workers.

The tour comprised of a visit to each of the three interconnected buildings of the plant—power train, body shop, and final assembly. It was led by Ed Killgore, a down-home, regular sort of guy dressed in cowboy boots and a plaid shirt, and possessing what appeared to be the keenest insight into every detail of auto production at Saturn from the most general to the very specific and technical, as evidenced by his expert handling of questions posed by the four people who accompanied me on the tour—three executives from German auto plants and a retired manager of the paint department in one of GM's more conventional plants.

From the comfortable distance provided by the swift-moving media train, the plant appeared to be quite advanced technically, with robots doing many of the chores typically done by workers only a few years ago. (I was hard pressed to find a human being anywhere in the body shop.) The plant also appeared to be a clean, safe, and healthy atmosphere for workers, and inordinately free from the watchful eyes of supervisors even in such places as final assembly. The groups of production workers that I saw during the tour seemed young (the average age, it turns out, is about thirty-six), ethnically diverse, and relatively unharried in their work.

At the completion of my tour, I met with Ms. Graham, a spunky and articulate Tennessee native, who escorted me, by my request, to final assembly. Over the course of the next couple of hours I spoke with roughly a dozen workers, but perhaps to only half that many in any depth. Workers were "on duty" during my entire visit to the shopfloor and so there was more than the occasional interruption. My questions focussed on workers' discontents with shopfloor conditions, especially the issue of pace which has plagued auto workers for nearly a century now, and the formal and informal institutions by which workers could voice their discontent with shopfloor conditions. The interviews were not taped, the workers with whom I spoke may not have been a representative sample, and the questions were not in the nature of a systematic survey. What follows, then, are my thoughts and impressions from these conversations.

Workers with whom I spoke were generally happy with the working conditions at Saturn, and happier, by far, when they compared these conditions with those they had experienced at other GM plants. Several workers explicitly mentioned the improvement in the quality of worklife made possible by the "skillet"—a moving, elevated wooden floor about fifteen feet wide on which workers stand to conduct their assembly operations instead of chasing a moving car down the assembly line, as in conventional plants. The skillet was one of Saturn's many production innovations. No complaints were voiced by workers on such issues as cleanliness, health and safety, or arbitrary and dictatorial behavior by superiors. (With a ratio of 50:1 in direct production of workers to indirect staff employees, compared to 25:1 at most GM plants, supervision is simply much less present at Saturn.)

With respect to work pace, while only a few workers failed to express the desire for it to be even more relaxed than at present, there was near unanimous agreement that the intensity of worker effort at Saturn was less than at other GM plants. Formal break periods—a half hour for lunch and two fifteen-minute breaks throughout the ten-hour day—do not offer much time for recuperation, but then again not all of one's time on the line is spent working. Team production at Saturn allows workers to decide collectively on the allocation of tasks, and thus the intensity with which each individual works and the informal break periods each consumes during normal production time. Many workers appear to be able to utilize this arrangement to carve out informal breaks throughout the work day.[1] (The first worker with whom I spoke was in the middle of a chess match with his buddy on one of the many wooden picnic tables that line the perimeter of the assembly line at Saturn. When he agreed to speak with me, his buddy—a fellow team member—took the opportunity to run off and check something on one of the regular tasks assigned to their team.)

Far more interesting than workers' answers to questions about shopfloor conditions were their responses to queries about the distribution of decision-making power over those conditions. When asked, for example, how the basic rudiments of work pace are formally determined, workers gave answers that were not very different from what one would hear in a conventional auto plant. Production standards and the package of tasks assigned to a team came down from the engineering department, and were established through procedures, such as time study, that involved some input from the union. Although the union's input in this case ostensibly stemmed from the unique partnering arrangement at Saturn, none of the workers explicitly mentioned the partnering system in any of our discussions of this issue.

Workers' answers were generally split when asked whether they had much say in this process. Some workers viewed decisions on such things as production standards as taking place somewhere far removed from the shopfloor, and largely beyond intervention by rank-and-file workers. The word "management" was used by these workers in describing people with decision-making authority, even though the formal partnering arrangement implies that some of these "managers" were fellow union members. These workers may be representatives of the contingent of the workforce— numbering, by recent union survey evidence, roughly 20 percent—that feels that the Saturn experiment is heading in the wrong direction (cited in Nauss 1994: D6). This is arguably the same group that has been behind efforts to win for workers the right to elect formally the worker representatives to the pairs of module advisors that oversee teams. A referendum in 1993 found 30 percent of the workforce in favor of making this change to the process of representation at Saturn (Rubinstein et al. 1993: 361).

Other workers with whom I spoke, however, felt that their ability to influence production matters was adequate. As one worker put it, in comparing the situation at Saturn with the other GM plant in which he worked, "If I have a problem, I can just call up engineering. I couldn't do that before." Access to engineering is made much easier by the fact that Saturn has relocated some of its engineering staff from Troy, Michigan, to the Spring Hill site of production. This transfer was initiated in part due to past tensions between engineering and work teams in which engineers saw little value in communicating with teams. It is perhaps noteworthy that this particular worker did not claim that his input was enhanced by the fact that he could speak with a union brother or sister in engineering when voicing his concern.

These two positions on the extent of worker empowerment reveal, despite their divergence, a common view of the nature of workers' influence in decision making at Saturn—namely, that it is based more on informal norms of labor–management behavior than on formal worker rights. One group of workers bemoans the lack of formal rights in shopfloor matters and strives to enhance its rights through, for example, demands for more democratic participation in the choice of worker representatives in management. The other group feels that its decision-making authority is adequate, but nonetheless admits that this authority rests on decidedly informal arrangements—e.g., "when I call up engineering, they listen." It is the ethos of the institution and informal norms of labor–management behavior that ensure a space for worker input at Saturn. In fact, formal worker rights are seen by management as anathema to labor–management cooperation, as a harkening back to the old days of adversarial collective bargaining.

Despite the absence of formal rights, the informal arrangements and ethos of the institution mean that workers' voices do carry significant weight. Nowhere is this clearer, perhaps, than in Saturn's cultural commitment to quality in production, and to the notion that workers are responsible for product quality. Even those workers who expressed disenchantment with their formal power in production admitted that being given responsibility for product quality has allowed workers to fend off some of management's "economizing" moves in the past. Workers may use their responsibility for quality to defend against management-initiated increases in work pace, but my conversations with workers in final assembly suggest that they have as much concern for the quality of the final product as for their own immediate shopfloor interests when assessing such decisions as the appropriate work pace or the proper packaging of tasks.

Past events at Saturn support this conclusion. Workers were influential in the very early stages of product and process design in convincing management to alter intended plans. Workers argued, for example, that the original design of the Saturn interior was too cramped and needed to be enlarged, a change to which management conceded somewhat late in the game and at no small expense (Sherman 1994: 236). Teams were also influential in convincing engineering to build the original prototype on the actual fixtures to be used in production, which revealed that the assembly could not be performed on the intended fixtures and allowed a debugging of the assembly process to occur hand-in-hand with product development (Rubinstein et al. 1993: 349).

Workers' ability to play the "quality" card has not been maintained without some struggle, however. In 1991, for example, work teams began wearing black arm bands that said "Stop Defects," and engaged in production slowdowns to protest what they viewed as an erosion of management's concern with quality, perhaps in reaction to brisk market demand that exceeded production volume at the time. Management was forced to respect this demonstration of concern, and responded by establishing a joint off-line problem-solving process between manufacturing, engineering, and operations management and by initiating the process of relocating some of the engineering staff from Michigan to Tennessee (Rubinstein et al. 1993: 355).

By most of the standards set out in the original conception, the Saturn project has been an enormous success. Saturn produces an affordable, high-quality compact car that consistently ranks in the top two or three—behind such luminaries as the Lexis and the Infinity—in terms of customer satisfaction (Aaker 1994: 116). While not quite co-determination, Saturn has created a truly unique worker empowering industrial relations environment all its own, based less on formal worker rights and more on informal behavioral norms.

From management's perspective, however, things do not look so good. The profit figures from Saturn have always been a sore spot for the company. Saturn has been only marginally profitable for the past few years, and even this has required some "creative accounting" in which development costs are partially forgiven (Nauss 1994: D6). Part of the explanation rests on the low profit margin of the compact vehicle market, but there is also the issue of worker productivity. The productivity statistics at Saturn have been a continuing source of disappointment to GM. Saturn reportedly compares unfavorably to other GM plants, where the work pace is greater. And compared to notable competitors such as the Honda Civic, the Saturn operation requires over 30 percent more worker-hours to produce a car (Nauss 1994: D6).

As a result, the Saturn project is apparently no longer viewed by GM management as the breeding ground for their new organizational approach to production. Recent evidence suggests that the company has chosen, instead, the more conventional lean production model at its NUMMI plant as the system of shopfloor governance for GM's future. Thus, if there is to be much transference of organizational methods across plants, it appears that Saturn will be the recipient rather than the initiator of future organizational changes.

Saturn's formal independence from the larger GM management structure was eliminated in 1994. Robert Boruff, Saturn's vice-president for manufacturing, has promised to significantly reduce the surplus in worker-hours per car that Saturn possesses in comparison with other automobile plants. The way forward is to increase the intensity of labor effort. GM has already intimated that it would like to reduce the number of assembly-line workers by 1,000 by the year 1997, while holding volume constant or even increasing it (Nauss 1994: D6).

The Saturn project reveals the benefits that are possible when workers' shopfloor power is significantly increased. It also reveals something about the likely distribution of those benefits. Customer and worker satisfaction from the Saturn project has been enormous, while corporate satisfaction has slowly waned. Apparently, the corporate share of the overall rewards from production using the Saturn approach compare unfavorably to that of the lean production model employed at NUMMI. While the total rewards to the various stakeholders from the cooperative venture at Saturn may well exceed those from the lean production approach, the latter is fast becoming the preferred choice by GM and, indeed, most of corporate America.

9

THE FUTURE OF US SHOPFLOOR GOVERNANCE

American industrial relations remain in a state of transition. For well over two decades now employers have engaged in experiments with workplace reorganization in reaction to the changing technologies of production, growing international competition, and smoldering discontent among workers over shopfloor governance. At present, these experiments appear to be relatively pervasive (Osterman 1994). They are also somewhat haphazard and incoherent. Appelbaum and Batt, arguably the single best source available on the nature and direction of recent workplace reorganizations, write:

> Our review of 185 consultants' reports and academic case studies of workplace change [lead us to the finding] that US companies have largely implemented innovations on a piece-meal basis and that most experiments do not add up to a coherent alternative [to prior practice].
>
> (Appelbaum and Batt 1994: 10)

Thus there appears to be an overwhelming sense among US employers that the older system of shopfloor governance needs replacing, and yet much ambivalence about what exactly to replace it with, suggesting that an opportunity still exists for coordinated efforts to affect the path of workplace reform.

Amidst the seemingly "piece-meal basis" of contemporary reform efforts, one can nonetheless identify the broad outlines of an emerging vision of industrial relations around which many employers seem to be coalescing—namely, a uniquely American version of the Japanese lean-production model. In this last chapter I argue that the lean-production approach to shopfloor governance is inferior, in terms of productive efficiency, to an alternative set of institutional arrangements that would grant workers greater participation rights in production. The lean-production model may offer limited improvements in certain areas of shopfloor governance, but it is not the best we can do.

The relative inferiority of the lean-production model stems from the fact that it fails to truly empower workers on the shopfloor. Worker shopfloor power affects the distribution of shopfloor rewards and the extent of managerial authority. Workers' sense of justice in distribution and the legitimacy of authority influences the extent to which they cooperate with management. Cooperation, in turn, affects productive efficiency.

Contemporary efforts to establish something similar to the Japanese lean-production model in the US will not foster the degree of cooperation between labor and management that is possible because the system fails to resolve the lingering problems of injustice and illegitimacy in shopfloor governance. However, combining certain aspects of the lean-production model with a set of statutory rights for workers through a works-council system, similar to that found in the German model of industrial relations, could help to achieve the labor–management cooperation we desire and the superior productive efficiency we need.

COOPERATION, PRODUCTIVE EFFICIENCY, SHOPFLOOR POWER, AND LEAN PRODUCTION

The extent to which labor and management cooperate in production is determined, in part, by the formal and informal institutional arrangements of shopfloor governance. Cooperation is fostered by improved communication between labor and management, an enhanced belief in the truth value of that communication, and by the belief on behalf of both parties that the structure of decision-making authority is legitimate and the distribution of rewards from production is just. Cooperation, in turn, fosters productive efficiency.

Productive efficiency is a measure of the joint welfare of stakeholders in the firm. It is not synonymous with productivity. Increased productivity may lead to reduced unit labor costs and increased profits, but also to increased intensity of labor effort and reduced workplace health and safety. Thus, productive efficiency cannot be measured by labor productivity alone, but rather requires a more complicated measure that takes into account shopfloor conditions as well—for example, output per unit of labor effort, controlling for the level of workplace health and safety.

Granting workers greater power in shopfloor decision making can enhance efficiency by making shopfloor outcomes more just and by rendering managerial authority more legitimate, thereby improving labor's willingness to cooperate with management.[1] Whether or not an improvement in efficiency can be obtained by granting workers greater power in production depends on the amount of power workers currently possess, and the amount that is required to sustain their sense of justice in the distribution of shopfloor rewards and the legitimacy of managerial authority. One of the contributions of our analysis of shopfloor contractualism is to suggest that workers' political and ethical concerns about the structure of shopfloor governance are important factors impeding labor–management cooperation today.[2] The results of our analysis of the American lean-production model suggest that it is incapable of fully resolving these shopfloor concerns.

176

If superior productive efficiency could be obtained by granting workers greater decision-making power in production, why is the US moving towards the adoption of the lean-production model, a relatively inefficient system of shopfloor governance? It is now commonly acknowledged that the technological and institutional arrangements of production may evolve in ways that are not productively efficient (Marglin 1974; Fairris 1995b; Freeman and Lazear 1995). New shopfloor institutions are introduced by employers with a concern for their effect on the control of and distribution of rewards from production, as well as their implications for productive efficiency. Even if productive efficiency is greatly enhanced by granting workers more shopfloor power, doing so also puts workers in a position to appropriate a larger share of the rewards from production, implying potential losses for employers. Workers' shopfloor empowerment might lead to improved safety, decreased intensity, and—despite improved output per unit of labor effort—perhaps decreased labor productivity and profits.

The allure of the lean-production model for US employers is that it appears to have fostered high levels of workplace cooperation from workers in Japan, and to have accomplished this without granting workers much substantive power through formal decision-making rights in production. Thus, lean production seemed to hold out the possibility of boosting labor–management cooperation while simultaneously preventing workers from garnering significant improvements in shopfloor conditions that threaten productivity and profits. That is, it seemed to represent an enticing combination of productive efficiency and distributional rewards for employers.

Evidence to date suggests that the experiments with lean production have not had the boost in labor–management cooperation and labor productivity that were perhaps expected. Some have argued that the failure of the US lean-production model is due to the unwillingnes of domestic employers to adopt the full panoply of supporting institutional arrangements associated with lean production in Japan.[3] While US firms envy certain features of the Japanese industrial relations system, they have been either unable or unwilling to institute other parts of the Japanese approach to industrial relations that operate in consort with these features, and which are arguably necessary for the significant labor–management cooperation observed in that country. Among these are such things as the virtual guarantee of lifetime employment which is granted to primary-sector workers in Japan; the large percentage, as much as a third in some cases, of the workers' compensation packages that is determined directly by company performance; and the significant commitment of Japanese firms to worker training.

The results of empirical studies lend some support to the claim that domestic efforts with lean production have failed due to their incomplete importation of the Japanese approach. Cooperation and productivity appear to be boosted much more significantly when quality circles and teams are combined with profit-sharing arrangements, employment security, and a long-run commitment by workers to learning new skills and increasing their responsibilities in production (e.g., Levine and Tyson 1990; Ichniowski *et al.* 1994).

However, the failure of American lean production is not due solely to the absence of these structural features of the Japanese lean-production model. Even if these structural features were added to the contemporary reform efforts of domestic employers, there would remain the lingering problem, from workers' perspectives, of injustice in shopfloor outcomes and the illegitimacy of managerial authority. Workers want a greater say in shopfloor governance, and the lean-production approach, even with these structural features added, grants few formal decision-making rights to workers (Marsh 1992; Lincoln and Kalleberg 1990).

At this historical juncture, the most productively efficient system of shopfloor governance would be one that grants workers greater power on the shopfloor. Following roughly three decades of declining shopfloor safety and experience with systems of shopfloor governance that are unresponsive to workers' shopfloor concerns, a shift in the distribution of power towards labor is required to restore justice, legitimacy, and efficiency in US production.

If the Japanese lean-production model is unable to generate the degree of labor–management cooperation, and thus productive efficiency, that is possible, we must look elsewhere for successful institutional arrangements of shopfloor governance after which to pattern our efforts at workplace reform. Germany possesses a system of industrial relations that, by most accounts, fosters significant labor–management cooperation; has yielded high rates of productivity growth and continuous improvements in workplace safety; and also possesses features, such as flexibility in labor allocation, which American management is currently striving to adopt. Because Germany's political system is a liberal democracy, in which industrial relations are based on principles of worker rights, it is also arguably more consistent with American political values.

THE GERMAN WORKS-COUNCIL SYSTEM

The legal framework governing industrial relations in Germany grants three spheres of influence to workers in production. The first is collective bargaining through unions, which is concerned primarily with the joint determination of wages and hours. The second is co-determination of firm policy by worker and management representatives on the governing boards of companies. And the third sphere of influence is the works-council system, which concerns itself primarily with worker consultation and joint determination of plant conditions with management.

The most important feature of the German industrial relations system, for our purposes, is the works-council system. Postwar legislation on works councils was inspired by related legislation during the Weimar Republic. The two primary Works Constitution Acts of the postwar period—one in

1952 and another in 1972—contain a number of important differences. One of these is the language governing a union's role in plant-level governance. The earlier act severely constrained the role of unions, banning them in effect as the shopfloor representatives of workers by granting this terrain to independent works councils. The later act altered this arrangement by, among other things, giving unions unlimited access to the plant with prior notification to management, allowing works councillors to be union officers, and allowing works-council representatives to engage in union activities on the shopfloor. Roughly 80 percent of works-council representatives in German factories and offices are union members (Markovits 1986: 49).

Works council legislation provides the opportunity for workers to elect a body to represent workers' interests in plant-level decision making. The creation of a works council can be initiated by either workers (at least three workers must begin the process) or the union, and results in the election of a council, the seats of which are filled by proportional representation from among various slates of candidates. Works councils represent both blue-collar and white-collar workers in proportion to their numbers in the plant.[4] Turnout in works-council elections is second only to those of general national elections in Germany, averaging around 80 percent of the plant workforce, with participation rates for blue-collar workers being higher than that for salaried employees (Muller-Jentsch 1995: 73). While no official statistics exist on works-council representation, Markovits cites estimates that place the concentration of works councils among eligible West German firms at only 18.9 percent, but covering 65.6 percent of the private-sector workforce (Markovits 1986: 48).

By legal mandate, the number of works-council representatives varies with the size of the establishment, from three representatives for establishments containing between twenty-one and fifty employees; fifteen for establishments between 1,000 and 2,000 employees; and thirty-five for establishments with 12,000 to 15,000 employees (Muller-Jentsch 1995: 58). The number of councillors released for full-time council activity is also legally mandated and varies with the establishment size, from three for establishments between 1,000 and 2,000 employees, to a minimum of fifteen for establishments between 12,000 and 15,000 employees (Muller-Jentsch 1995: 58). Council members currently serve a term of office of four years, up from the three-year terms which existed prior to 1989 and the two-year terms of office that existed prior to 1972 (Muller-Jentsch 1995: 57).

Works councils are generally thought of as establishment-level worker organizations. However, multi-plant firms in Germany typically have central works councils, composed of the delegates from the various establishment-level councils. And more broadly encompassing works councils can be created across groups of firms if works-council delegates representing at least 75 percent of the workforce of a group of firms request that a council be established (Muller-Jentsch 1995: 55).

179

Works councils meet at least once a month with the employer. Works council representatives also engage in the resolution of worker grievances on a day-to-day basis. The employer bears all of the costs associated with council activities, including council elections, conciliation committees, the salaries of councillors who are released for full-time council business, and staff support and supplies. In large companies, works councils typically have an office and a staff, sometimes including expert staff people for advice on special matters of concern to the plant workforce (Muller-Jentsch 1995: 65).

Works councils are granted certain rights and are saddled with certain obligations. The chief obligation of works councils is to "cooperate" with the management, union, and employers' association in "a spirit of mutual trust" for "the good of the plant and its workers" (part of the language of the act, cited in Markovits 1986: 49 and Muller-Jentsch 1995: 58). They are obligated to uphold and monitor the collective-bargaining agreement reached by the union and employers' association, any social legislation (e.g., safety regulations) that may apply to the plant's operation, and any internal agreements reached between the works council and plant management. They also serve as the mechanism for resolving worker grievances with management. The obligation of works councils to "cooperate" is given legal reinforcement in that they are prevented from using strikes to settle grievances; conflicts are formally resolved (in those cases where joint determination is either legally or customarily required) through arbitration proceedings with boards composed of an equal number of works council and company representatives and a neutral chair.

In return for these obligations, works councils are granted certain statutory rights in production decisions. In some cases this is merely the right to be consulted by management, but in other cases works councils possess rights of joint determination with management over plant-level decisions. Works councils can, in addition, sign agreements with management that have legal force similar to collective-bargaining agreements, thereby limiting managerial prerogative over the range of decisions for which they do not possess statutory rights of decision making.[5] Collective bargaining agreements always take precedence over any local works-council agreement, however, and local agreements are, in contrast to union contracts, much easier for management to circumvent should conditions necessitate (Markovits 1986: 50).

Formally, works councils have the statutory right of joint determination over the structure of the workday (e.g., starting and quitting times, overtime, and the spacing of break periods); those aspects of pay which are only vaguely spelled out in the collective-bargaining agreement (e.g., piece rates and the methods by which they are calculated); decisions concerning vocational training; certain personnel practices (e.g., establishing criteria for hiring, firing, and transfer of workers); safety ("arrangements for the prevention of employment accidents and occupational diseases, and for the protection of health on the basis of legislation or safety regulations" (Muller-Jentsch 1995: 59); and the introduction and use of technical devices designed to monitor worker performance (Thimm 1980: 38; Muller-Jentsch 1995: 59). Management is legally prevented from making unilateral decisions on these matters. Disputes concerning such matters, as well as any disputes associated with collective-bargaining agreements or local plant accords, are subject to arbitration.

There is, in addition, a range of decisions over which works councils are granted more limited rights of joint determination. These decisions involve structural changes such as plant closings or transfer of work to other facilities, mergers, significant changes in the plant's organization or product, and major reorganization in work methods (Thimm 1980: 38). Management is compelled, in such cases, to consult with the works council and to establish an implementation plan that is fair to the workforce of the plant. Mediation can be solicited by the plant arbitration board or by the labor court in the event of unresolveable disagreement, but in the end management can act unilaterally if no resolution is reached.

Finally, there are plant decisions over which management is required only to consult with works councils before implementation. Such issues as technological changes, job redesign, and mass layoffs fall into this category. For that range of decisions over which works councils are not granted joint determinative power with management, the process of consultation and, in some cases, mediation of disagreements can provide workers with a significant amount of leverage. Moreover, works councils often "package" different subjects for discussion with management in such a way that the works council's stance on jointly determined issues is contingent upon management's decision with respect to other issues that require only consultation with workers (Streeck 1984: 25).

The inability to strike removes a certain amount of leverage that works councils might otherwise possess with regard to managerial decisions. But, given their statutory rights of joint determination and consultation over a wide range of shopfloor decisions, works councils are well poised to disrupt production through foot-dragging or lack of cooperation over routine matters of day-to-day production. Management must elicit the consent of workers on a range of shopfloor issues, whether works councils are granted joint-determinative rights or not, in order for production to run smoothly.[6]

There is a division of labor of sorts in the dual system of worker representation in Germany, with unions bargaining over wages, hours, and benefits, and works councils influencing workplace conditions. However, the division of duties is by no means perfect. Because collective bargaining takes place at the industry or—in the case of metals, for example—the multi-industry level, and is typically conducted regionally with significant national coordination, unions and employer associations are forced to agree on the basic wage structure using only very broadly defined skill categories.[7] Works councils possess a number of outlets for further influencing worker pay levels once these broadly drawn categories are agreed upon.

By the same token, while working conditions are largely the terrain of works councils, unions can influence such conditions by including manning requirements for specific jobs or even more detailed stipulations about working conditions for particular skill groups in the appendices to contracts (Markovits 1986: 42). Unions can also negotiate general rules governing working conditions in industry-wide agreements, with the further contractual stipulation that works councils be mandated to formulate more specific language governing such conditions in supplementary "works agreements" (Muller-Jentsch 1995: 62).

This reference to a division of duties between unions and works councils is not meant to suggest that the two always act jointly, in a coordinated and cooperative fashion, to represent workers interests in these two different realms of worker concern. In the early postwar period there was much independence of action between the two organizations. While this independence has lessened over time—so that many German trade union officials currently consider works councils to be "the extended arm of the union" (Streeck 1984: 27)—formal independence still remains, and this independence gives works councils a certain amount of necessary leverage in shaping union goals. German trade unions are powerful, bureaucratic organizations, and, like unions in the US, their concern with workers' shopfloor interests must be constantly nourished. Works councils help in this process because they are functionally useful for unions. German works councils are legally responsible for policing management's adherence to the terms of the collective-bargaining agreement, for example, and, in addition, they can act as a union recruitment mechanism in the shop—something that is extremely useful given that, by law, union membership cannot be made compulsory. Thus, because works councils possess independent power through statutory rights and are also in a position to garner benefits for unions, they are able to elicit union cooperation in achieving workers' shopfloor goals (Rogers and Streeck 1994).

To see how works councils can represent workers' shopfloor concerns, it is illustrative to compare the responses of firms, workers, unions, and government officials in the US and Germany to the pressures placed on the manufacturing sectors of both countries during the 1950s and 1960s. We have covered the US case in earlier chapters. Markovits (1986) notes that increased international competition in steel during the early 1960s spurred German steel manufacturers to intensify production, resulting in a deterioration in shopfloor conditions. German manufacturing faced its first serious postwar recession in 1966–7, roughly a decade after American manufacturing, thereby placing similar pressures on a wide variety of industries.

In Germany, in contrast to the US, a crisis of legitimacy in managerial authority and worker feelings of injustice in the distribution of shopfloor rewards appear to have been avoided. In some areas, works councils possessed sufficient rights of shopfloor influence to successfully repel efforts by employers to unilaterally raise production standards and reduce the quality of shopfloor conditions. In areas where worker rights were inadequate, workers were able, through the support of unions, to enhance their shopfloor power *vis-à-vis* management.

Workers reacted strongly to the pressures placed on them by employers during the 1960s.[8] Spontaneous strikes emerged in a handful of plants in the early recession years, but grew in number and intensity during the recovery, reaching a peak in 1969. One of the workers' demands during this conflict was for greater "on-the-job codetermination."[9] This was presumably an attempt by workers to stem the tide of employer efforts to reduce the quality of shopfloor conditions. Unions were persuaded to back this rank-and-file demand, and there emerged strong union support for a variety of measures to bolster the quality of worklife—among the most important being efforts to extend the shopfloor rights of works councils by revisions to the Works Constitution Act of 1952.

In 1972, the revised Act extended works-council rights of co-determination over various aspects of the hours and timing of work, as well as the payment system. It extended rights of consultation to works councils over plant changes in working conditions due to the adoption of new technology or alterations in the organizational design, and created significant rights to ensure that workers' interests would be considered in the event of mass layoffs or plant closure. The Act also allowed for greater integration of works council and union activity in the plant.

The role of works councils in workplace health and safety was also enhanced during this period, in part by provisions of the 1972 Act, but also through the introduction of new industry regulations (with which works councils are empowered to force management compliance). German workers possess rights to information about health and safety hazards, rights to monitor management compliance with regulations, and the ability to veto or withhold consent to certain management decisions concerning heath and safety (Gevers 1983).[10]

Works councils have remained equally important institutions for representing workers' interests in the 1980s and 1990s. In part, this is because the shopfloor has been such an important component in the process of adjustment to new technologies and increased competition in international markets, and works councils are largely responsible for this terrain. But it is also the case that employers have resisted other avenues of influence, thereby elevating the status of works councils. Works councils are uniquely situated to consider both worker and company interests in technological and organizational changes in production. They cooperate with management because they are sufficiently empowered to ensure the legitimacy of the process and the fairness of the outcome of work reorganization.

In the early 1980s, IG Metall—the union representing workers in the metal industry in Germany—introduced a significant change in its policy towards the introduction of new technologies. The central union would establish general principles, offer technical advice, and coordinate efforts across plants by sharing more widely the specific experiences of certain successful works council efforts, but henceforth works councils and local union organizations would be given much greater responsibility for responding to plant-level technological change. For example, works councils would be encouraged to form special technology committees and shop stewards would be involved in efforts to develop alternative plans for work reorganization to bring into negotiations with management.

The Works Constitution Act was amended again in 1989 to address the issue of technological change. Labor wanted full co-determination rights over the introduction of new technology. While the new law fell short of full co-determination rights for works councils, it strengthened already existing consultation and information rights, mandating, for example, that an employer consult with its works council far enough in advance of a change in the technology of production so that suggestions and objections could be fully considered; that affected employees be informed of the company's plans; and that discussions ensue concerning how these workers' skills could be adapted to future requirements (Muller-Jentsch 1995: 59).

It was also during this period that unions and works councils were confronted with companies' growing interest in quality circles and team production. Initially, the union's position was that these posed a threat to the representation of workers' interests in shopfloor decisions, and the advice to works councils was to refuse to participate. More recently, however, the policy has become more nuanced. Since the mid-1980s, works agreements between works councils and employers have emerged in many industries specifying the composition of quality circles, their procedures, and the role of works councils in their governance. Similar agreements have emerged around the issue of teams, specifying, for example, the election of the team leader and greater time for the discussion of production matters in team meetings (Muller-Jentsch 1995: 69–70). German unions and works councils have done much to ensure that shopfloor participation schemes further empower workers in shopfloor decision making.

In autos, for example, the union has recently initiated proposals for work reorganization involving greater flexibility in production that offer a distinct alternative to the lean-production approach of the Japanese model. Known as "group work" (Turner 1991), the union proposal represents ideas and experiences that emerged from the rank-and-file unrest of the late 1960s and the resulting experiments with work humanization in the early 1970s. They include, among other things, job rotation among tasks ; longer cycle times; autonomy of group decision making in the areas of quality control, training requirements, and division of the work; the adoption of technologies consistent with group work; equal pay for group members; and a joint steering committee at the firm level composed of equal numbers of workers and management to oversee its implementation (Turner 1991: 114).[11] The final form group work might take if implemented more generally in autos (and, in particular, how distinct it may truly be from "team production" in the US) is still very much an open question.

WORKS COUNCILS AND THE FUTURE OF US SHOPFLOOR GOVERNANCE

US workplace reform efforts should include measures granting workers greater participation rights in production. Doing so would address workers' existing concerns with injustice in the distribution of shopfloor rewards and illegitimacy in the authority possessed by management, thereby fostering greater worker cooperation in production and thus superior productive efficiency. A system of statutory works-council rights could serve as a model for accomplishing this goal. The adoption of a works-council system would confer certain benefits on each of the various stakeholders in production.

The shopfloor benefits for workers from a works-council system are clear: a collective-voice mechanism for those workers who currently lack one; a greatly expanded range of jointly-determined shopfloor decisions; continuous negotiations between labor and management over changing shopfloor conditions (as opposed to periodic contract negotiations or none at all); increased leverage over unions to focus their attention on workers' shopfloor concerns; and a set of inalienable rights of participation in production.

The union movement also stands to benefit from the introduction of a works-council system in the US. Typically, unions are skeptical of efforts to grant workers greater autonomy at the plant level, fearing that this autonomy will lead either to too much cooperation with management or to too much militancy (Rogers and Streeck 1995). The former militates against union formation while the latter can be a destabilizing force for existing union structures.

However, there are reasons to believe that a works-council system in the US today would encourage, rather than discourage, the expansion of unions and collective bargaining, just as was true in the early twentieth century with company unions. Demands for representation, once granted over a limited set of issues, would grow to encompass other issues best handled by unions. Moreover, the staff and expertise that national and international unions are able to offer workers on the range of issues of concern to them would probably greatly exceed those of works councils, adding yet another incentive for workers to affiliate with a union (Adams 1985).

In addition to workers and unions, a works-council system might benefit society more generally by lowering the costs of achieving socially desirable employment outcomes which are currently the responsibility of government regulatory agencies to oversee and enforce (Weiler 1990). Government regulation of health and safety in the US workplace is a good example. Federal safety policy in the US has taken a standards-setting approach, in which safety standards are set and federal regulators inspect workplaces to ensure compliance with those standards. The standards-setting process is very slow; after two decades of regulation, existing standards cover, at best, 25 percent of the industrial accidents in the US. Thus, bringing employers into full compliance with regulatory standards would reduce workplace injuries by, at most, 25 percent. Compliance is an additional problem. Most plants in the US have less than a 10 percent chance of being inspected by government inspectors in any given year, and, moreover, even if inspected and found to be in violation of a federal safety standard, the average fine per serious violation is less than $300 (Smith 1992: 567).

The slowness of the standards-setting process and the inadequacy of the compliance procedures are, to a certain extent, a manifestation of the real problem with this approach to safety regulation—namely, that it is virtually impossible to put forth permanent and homogeneous safety standards for dynamically changing and heterogeneous workplaces, and to have these standards adequately enforced by outside agents. Properly empowered works councils might provide the basis for greater flexibility in standards of health and safety across diverse plants, firms, and industries, as well as a far superior enforcement mechanism.

There are potential benefits from a works-council system for employers as well. The German system possesses many of the features US employers seem to want in a new set of institutional arrangements for shopfloor governance. For example, in German manufacturing there is typically greater flexibility in the utilization of labor resources and workers are given greater responsibility for troubleshooting and quality control in their work. Works councils foster such outcomes by elevating workers' rights and responsibilities in production, making workers less fearful of the negative consequences of flexibility and more willing to embrace the goal of high-quality production.

Moreover, works councils are not a substitute for, as much as they are a complement to, emerging forms of joint communication committees and work teams in production, which have recently become so pervasive in US firms. A works-council system acts as a collective-voice mechanism—similar in spirit to company unions but with more power—that can coordinate the disparate and sometimes competing interests of work teams, and collectivize worker preferences over those shopfloor conditions that constitute workplace public goods. This serves the interest of employers as well as workers.[12]

Although a works-council system might benefit employers in certain respects, our analysis of employers' reactions to the system of fractional bargaining and the contemporary Saturn project suggests that a system of shopfloor governance that grants workers great shopfloor power, and thereby significant authority in production and improved shopfloor conditions, is not likely to receive much lasting support from employers. American

186

employers have historically opposed any measures that threaten the sanctity of managerial prerogatives in production, and are likely to do so today as well. This not only suggests that a costly and divisive battle is likely to ensue over the creation of a works-council system in the US, it also raises concerns about the degree to which management would view such a system as legitimate and fair, and thus whether such a system would elicit management's cooperation with workers if successfully established.

One response to these concerns is to argue that employers' reactions to the introduction of works councils in the US might well be similar to their experiences during World War I when forced to adopt worker voice mechanisms for dispute resolution by the War Labor Board. Initially, employers were very reluctant to grant a voice to workers, but learned from the experience that mechanisms for collective worker voice served certain interests of management as well. German managers routinely express approval of works councils for their role as a mechanism for information sharing and the representation of interest groups amongst the workers (e.g., Muller-Jentsch 1995; Freeman and Lazear 1995).

But this response may be too optimistic. Employers seem more and more committed to the American lean-production model, and the reason for this is not its productive efficiency but rather its implications for the distribution of shopfloor rewards and shopfloor authority. Thus, perhaps we must simply admit that fostering greater productive efficiency through a works-council system will threaten the power that employers possess in, and the rewards they receive from, production. Acknowledging this, the best we can do is to move forward in a democratic process of open communication and good will, searching for ways, when possible, to share some of the enhanced rewards from superior productive efficiency with employers.

In the end, of course, the degree to which employers are threatened by a works-council system will depend on the specific rights workers are granted by virtue of its creation. There are many important issues to be settled. How extensive are workers' rights of consultation and co-determination to be? Who is responsible for the financial resources of the works council, employers or workers, and can unions contribute to these financial resources? What procedures—for example, binding arbitration or the right to strike—should be put in place for dispute resolution? What measures are necessary to ensure due process and fair representation for workers in dispute resolution procedures?

The answers to many of these questions are best left to the debates that will unfold if we embark on the path of creating a works-council system in the US. At this historical juncture, however, given the illegitimacy of managerial authority and injustice in the distribution of shopfloor rewards that workers currently perceive, a minimal set of worker rights sufficient to foster significant improvements in productive efficiency would seem to include the following:[13]

1 Works councils would be granted an extensive set of rights to information on and *consultation* about a wide range of firm policies affecting the shopfloor, including job descriptions, production standards, worker training, procedures for the internal allocation of labor, and proposed changes in the technology and organization of production.
2 Works councils would become responsible for the local administration of laws governing shopfloor conditions, including, most obviously, workplace health and safety.
3 Works councils would be granted co-determination rights over the health and safety policies of firms. This would include full information and inspection rights, as well as the right to stop production when warranted by hazardous conditions.[14]

Thus far we have discussed works councils in isolation from other formal and informal institutional arrangements. A close analysis of the German case suggests that there are an array of formal institutions—such as the apprenticeship system of worker training and powerful national unions with centralized bargaining structures—that in one way or another support the operation of works councils, and are arguably important for the performance outcomes works councils generate in that country. It remains to be seen how works councils will function in a country like the US where supporting institutions such as these are absent.

Certain informal institutional arrangements may be equally important. Our analysis of the change in shopfloor governance during the late 1930s suggests that formal worker rights, whether those contained in the National Labor Relations Act or in the collective-bargaining agreements of the period, were by no means sufficient for the attainment of workers' shopfloor empowerment. Much of that empowerment was attained through the building of informal shopfloor organizations to aid in the union-organizing drives of the period. Works-council legislation would grant workers certain rights of participation on the shopfloor, but the empowerment workers attain through such rights is likely to be significantly related to the informal shopfloor institutions they build prior to and following works-council legislation.

For this reason, movements among workers to struggle against the confining aspects of lean-production methods should be encouraged. Unfortunately, organized opposition to lean production has often left the impression that a return to the older system of shopfloor contractualism is the preferred outcome. Given the significantly greater deterioration in shopfloor conditions in lean-production settings, it is perhaps understandable that some workers would wish to return to earlier times. But returning to the system of shopfloor contractualism is not a step forward. Instead, workers should take seriously employers' rhetoric of worker participation, and organize to turn back the negative consequences of lean production while embracing the opportunity for participation and greater responsibility in production.

When workers in lean-production plants demand the right to elect their team leaders, they should be encouraged. When workers organize to achieve improvements in health and safety, and a greater voice in the local administration of company health and safety policy, they should be applauded. When workers struggle against management speedup that leads to worker stress and to lower quality products, they should be seen as heroes. And when workers form to make their greater responsibility in production the basis for greater worker cooperation within and across informal work groups (as well as with management), they should receive our support.

Enhanced worker empowerment in production will help to restore legitimacy to managerial authority and justice to the distribution of shopfloor rewards. It will enhance workers' cooperation with management in production, and boost productive efficiency. Its impact on labor productivity, however, remains an open question. Less intensity and greater safety may be accomplished only with lower productivity. Looking for ways to boost worker productivity, while safeguarding shopfloor conditions, will be one of the joint challenges facing labor and management under a system of works-council rights.

Works councils, in combination with truly empowered informal worker shopfloor organizations, may thus provide both the impetus for and worker voice to accomplish the movement to a "high performance" workplace (about which we hear so much these days), in which workers' skills are significantly increased through training, and their responsibilities in production are expanded accordingly. One of the lessons to be drawn from the German experience with works councils is that a highly participatory work environment for workers is compatible, not with a de-skilled, dominated, and demoralized workforce, but with workers who are highly skilled, autonomous, and encouraged to take responsibility for their own welfare, as well as the long-run viability of the firm for which they work.

CONCLUSION

The twentieth-century history of shopfloor governance in American manufacturing has been a story of great progress followed by significant stagnation and even decline. The first fifty years or so witnessed rising worker empowerment on the shopfloor. Workers attained the ability to voice their shopfloor concerns to management through company unions and then the power to force management to accede to these concerns through unionization and the informal shopfloor organizations that became part of early union structures. Productivity growth during this period was unprecedented. Workplace health and safety witnessed steady improvements, as did other shopfloor conditions such as the pace of production and the respect workers commanded from foremen and supervisors.

The last third of the century has not proven to be as beneficial for either labor or management. Shortly after mid-century, employers embarked on an effort to reduce the power of workers' informal shopfloor organizations by limiting workers' shopfloor rights to those formally spelled out in the language of collective-bargaining agreements. Workers' shopfloor power deteriorated, along with safety and other shopfloor conditions, but so too did their sense of justice and legitimacy in shopfloor governance, and thus their willingness to cooperate with management in production. For nearly two decades, between the mid-1960s and mid-1980s, productivity growth was anemic.

Other factors can account for some of the success of the earlier period as well as some of the difficulties of the later years. Among these, the economy's ability to adjust to changes in the technology of production and to changes in international markets stands out as especially important. But the evolution of shopfloor institutional arrangements has mattered as well.

In reaction to the slow productivity growth and rising worker discontent of the later period, employers have adopted new institutional arrangements that further centralize shopfloor decision making, but decentralize shopfloor dispute resolution and invite workers to become more responsible for troubleshooting and product quality. The emerging lean-production model fails, however, to garner the cooperation from workers that is possible because it does little to resolve workers' feelings of injustice and illegitimacy in the system of shopfloor governance of the last third of the century.

As a new century dawns, it is important that we realize there is a better way. A system of shopfloor governance that grants workers greater rights of decision making on the shopfloor would restore workers' sense of justice in the distribution of shopfloor rewards and legitimacy in the shopfloor authority of management. Combining aspects of the lean-production model—teams and greater responsibility for workers in production—with statutory rights of worker participation, could provide us with improved shopfloor conditions, boosted labor–management cooperation, and superior productive performance with which to start the twenty-first century.

APPENDIX TABLES

Appendix table 1.1

Regression equations: means (standard deviations) and estimated coefficients (standard errors)

*(1) $P_{1921-25}$ = 2.68 + 8.06 • Company union $_{1923}$
13.93 (7.04) (2.4) 1.39
(24.15) (1.96)

(2) $P_{1921-25}$ = -0.52 - 0.009 • Company union $_{1923}$ - 0.57 • $I_{1921-25}$
(6.53) (4.26) (0.23) -25.53
(36.85)

(3) $I_{1921-25}$ = -5.64 - 14.25 • Company union $_{1923}$
-25.53 (11.56) (5.02)
(36.85)

(4) $I_{1921-25}$ = 52.53 - 36.53 • Company union $_{1923}$ - 2.82 • HP/W $_{1921-25}$ + 1.42 • Company union • HP/W
(33.00) (12.15) (1.43) 23.32 (0.77) 27.30
(8.85) (29.02)

Nonparametric Measure of Ranking Correlation

N_C	N_D	S	τ	N
20	8	12	0.43	8

*Indicates correction for heteroskedasticity

Appendix table 1.2

Regression equations

(1) $P_{1925-29} = 16.85 + 4.19 \cdot \text{Company union}_{1924}$
  22.87   (7.06) (2.56)
  (22.42)   1.44
                            (2.46)

(2) $P_{1925-29} = 11.78 + 4.08 \cdot \text{Company union}_{1924} - 0.37 \cdot I_{1925-28}$
              (6.72) (2.25)   (0.19)   -14.15
                                                             (29.66)

*(3) $I_{1925-28} = -13.72 - 0.3 \cdot \text{Company union}_{1924}$
  -14.15   (9.5) (3.41)
  (29.66)   1.44
                            (2.46)

(4) $I_{1925-28} = -5.57 - 10.59 \cdot \text{Company union}_{1924} - 0.64 \cdot \text{HP/W}_{1925-29} + 0.4 \cdot \text{Company Union} \cdot \text{HP/W}$
              (16.95) (7.1)   (1.23)   15.51   0.3   42.25
                                                              17.00   (99.42)

Nonparametric Measure of Ranking Correlation

N_C	N_D	S	τ	N
43	23	20	0.30	12

*Indicates correction for heteroskedasticity

Appendix table 3.1

Regression equations

(1)	$I_{1936-40}$ -12.41 (15.90)	=	-10.21 (7.96)	−	1.05 (3.47)	•	$Union_{1941}$ 2.08 (0.97)		
(2)	$P_{1937-39}$ 5.94 (7.81)	=	18.10 (3.35)	−	6.32 (1.57)	•	$Union_{1941}$ 1.92 (0.95)		
(3)	P_{1937-9}	=	15.84 (3.70)	−	5.76 (1.59)	•	$Union_{1941}$ − 9.92 (7.75)	•	$I_{1937-39}$ -0.12 (0.20)

Regression equations

(1) $I_{1948\text{-}60}$ [−2.09 (2.02)] = −0.62 (2.13) + 0.02 (0.04) • $UNION_{1953}$ [45.26 (12.45)] − 0.88 (0.53) • SP [2.66 (0.88)]

**(2)   U–shape [0.65 (0.49)] = −1.26 (1.7) + 0.0008 (0.03) • $UNION_{1958}$ [49.53 (11.93)] + 0.63 (0.39) • SP [2.56 (0.86)]

(3) $P_{1948\text{-}60}$ [2.03 (0.91)] = 2.52 (1.0) + 0.006 (0.02) • $UNION_{1953}$ [45.26 (12.45)] − 0.30 (0.25) • SP [2.66 (0.88)]

(4) $P_{1948\text{-}60}$ = 2.45 (1.02) + 0.008 (0.02) • $UNION_{1953}$ − 0.36 (0.27) • SP − 0.07 (0.12) • $I_{1948\text{-}60}$ [−2.09 (2.02)]

*(5)   PI [0.89 (0.32)] = 0.73 (0.25) + 0.007 (0.005) • $UNION_{1958}$ + 0.18 (0.09) • SP

*(6)   PD [0.89 (0.32)] = 0.83 (0.20) − 0.007 (0.004) • $UNION_{1968}$ + 0.18 (0.08) • SP

*Indicates correction for heteroskedasticity

**Indicates probit estimation

Appendix table 7.1

Regression equations

$$(1)\ I_{1986-91} = -0.57 + 0.006 \cdot QC - 0.008 \cdot TEAMS + 0.01 \cdot TQM + 0.005 \cdot ROTATION$$
$$\qquad\qquad\quad (1.01)\ \ (0.009)\qquad\quad (0.008)\qquad\qquad (0.007)\qquad\quad (0.007)$$
$$+ 0.008 \cdot \%UNION - 0.0007 \cdot PLANT\ SIZE - 0.17E\text{-}07 \cdot PSIZE^2 - 0.02 \cdot \%FEMALE$$
$$(0.006)\qquad\qquad (0.0002)\qquad\qquad\quad (0.65E\text{-}08)\qquad\quad (0.01)$$
$$+ 0.05 \cdot \%B.COLLAR + 0.05 \cdot YEARS\ EXPERIENCE$$
$$(0.01)\qquad\qquad\quad (0.07)$$

Variable means (standard deviations):

Variable	Mean	(S.D.)
$I_{1986-91}$	3.93	(4.93)
QC	37.80	(39.27)
TEAMS	30.04	(39.08)
TQM	41.97	(44.80)
ROTATION	32.0	(38.03)
%UNION	50.46	(48.53)
PLANT SIZE	1862	(3301)
$PSIZE^2$	1432.8E+04	(8342.2E+04)
%FEMALE	32.44	(28.52)
%B.COLLAR	64.52	(23.00)
YEARS EXPERIENCE	4.2	(4.4)

$$(2)\ I_{1986-91} = -1.27 + 0.005 \cdot QC - 0.007 \cdot TEAMS + 0.01 \cdot TQM + 0.003 \cdot ROTATION$$
$$\qquad\qquad\quad (1.06)\ \ (0.009)\qquad\quad (0.008)\qquad\qquad (0.007)\qquad\quad (0.007)$$
$$+ 0.007 \cdot \%UNION - 0.0007 \cdot PLANT\ SIZE - 0.17E\text{-}07 \cdot PSIZE^2 - 0.02 \cdot \%FEMALE$$
$$(0.006)\qquad\qquad (0.0002)\qquad\qquad\quad (0.65E\text{-}08)\qquad\quad (0.01)$$
$$+ 0.05 \cdot \%B.COLLAR + 0.02 \cdot YEARS\ EXPERIENCE + 1.19 \cdot \Delta SKILL$$
$$(0.01)\qquad\qquad\quad (0.07)\qquad\qquad\qquad\qquad (0.57)$$

$\Delta SKILL$ mean 0.64 (0.48)

(3)

$CTDs_{1986-91}$ = 37.72 + 0.14 • QC − 0.25 • TEAMS − 0.003 • TQM + 0.01 • ROTATION
65.50 .... (7.92) .... (0.09) .... 37.54 .... (0.09) .... 30.40 .... (0.08) .... 41.29 .... (0.07) 32.0
(54.01) .... (39.39) .... (39.07) .... (44.67) .... (37.79)

− 0.07 • %UNION + 0.006 • PLANT SIZE − 0.13E-06 • $PLANT\ SIZE^2$ + 0.07 • %FEMALE
(0.08) .... 50.37 .... (0.003) .... 1853 .... (0.8E-07) .... $1431E+04$ .... (0.08) 32.62
.... (48.47) .... (3304) .... (8342E+04) 28.75

+ 0.34 • B.COLLAR + 1.11 • YEARS EXPERIENCE
(0.12) .... 63.95 .... (1.04) 4.1
.... (23.44) .... (4.4)

(4)

$CTDs_{1986-91}$ = 34.71 + 0.13 • QC − 0.24 • TEAMS − 0.005 • TQM + 0.006 • ROTATION
(8.47) .... (0.09) .... (0.09) .... (0.08) .... (0.07)

− 0.07 • %UNION + 0.006 • PLANT SIZE − 0.13E-06 • $PLANT\ SIZE^2$ + 0.07 • %FEMALE
(0.08) .... (0.003) .... (0.8E-07) .... (0.08)

+ 0.35 • B.COLLAR + 1.02 • YEARS EXPERIENCE + 5.26 • ΔSKILL
(0.12) .... (1.02) .... (5.39) .... (0.64)
.... (0.48)

*Indicates correction for heteroskedasticity

DATA APPENDIX

The data for the analyses preceding the presentation of Table 1.1 results come from the following sources: separation rates come from two field investigations conducted by the Bureau of Labor Statistics (one in 1914 and another in 1918, both investigating turnover during a twelve-month period ending roughly in the middle of the survey year) covering nearly 260 establishments, and employing over 500,000 workers (Brissenden and Frankel 1920: 1,350); union growth rates are from Wolman's analysis of union records, including principally, but not exclusively, the American Federation of Labor's annual report on membership in each of its affiliates (Wolman 1924: 88); strike data are compiled from two Bureau of Labor Statistics studies (Peterson 1937: 38; US Department of Labor 1917: 604), and derive from follow-up surveys of firms and unions reported to have been involved in a strike by a newspaper, labor or trade journal, or other printed source; and the data on welfare measures, their administration and impact, are from a 1916–17 Bureau of Labor Statistics investigation of 431 establishments, employing over 1,500,000 workers (US Department of Labor 1919: 15, 34, 43, 54, 70, 82, 89, and 119).

Many problems exist with the data. None of the surveys claims to be representative samples of American industry. Perhaps more importantly, there may well be significant inconsistency in the industrial classifications. Because no well-defined scheme of industrial classification existed at this time, it is quite possible for a particular category—for example, iron and steel—to embrace different sets of firms in different surveys, sampling differences aside. In one instance an inconsistency was imposed by the author: the data on welfare measures in the "food products" industry are joined with separation rates and strikes in "slaughtering and meat packing" to form what I have labeled the "food products" industry.

The precise definitions of the variables used in the analyses are as follows: *separation rate* is the percentage change in quits, discharges, and layoffs per 10,000 labor hours between 1913 and 1914 and between 1917 and 1918; *strike growth* is the percentage change in the number of strikes from 1914 to the average for the period 1915–20 (in four of the eight industries, the base year 1914 was unavailable, so 1915 was used instead for comparison with the average for 1916–20); *union growth* is the percentage change between 1910 and 1920 in the percent of the labor force organized in each industry; *welfare index* is the average of the percentage of establishments in each industry (a) possessing a hospital or emergency rooms, (b) granting rest periods to workers, (c) having wash rooms or showers, or both, (d) having restaurants, cafeterias, or lunch rooms, (e) possessing rest and recreation rooms, (f) having organized social gatherings—lectures, music, etc., and (g) having outdoor recreation facilities; *reduced turnover* is the percentage of establishments in each industry reporting that welfare measures improved the "stability of the (labor) force"; and *joint administration* is the percentage of establishments in each industry reporting that welfare work is jointly administered by the employer and employees.

The data for the analysis in Table 1.1 come from the following sources: the number of establishments with company unions by industry is taken from the National Industrial Conference Board's (1925: 8, Table 3) census of company unions for the years 1922 and 1924. (The average number of company unions for these two years was used.) The concentration of company unions by industry was then created using the number of establishments by manufacturing industry for 1923 from the *Historical Statistics of the United States: Colonial Times to 1970* (US Bureau of the Census 1975: 669–80). The data on growth in productivity rates and injury rates in selected industries come from a 1926 survey of establishments conducted by the American Engineering Council (1928) at the request of the National Bureau of Casualty and Surety Underwriters, who were very much interested in the impact on productivity of improvements in safety conditions. Data on horsepower and number of wage earners are from the *Statistical Abstract of the United States 1926* (US Department of Commerce 1926: 748–73).

Every industry that produced a match across the three data sets was used (with the exception of printing/publishing and clothing, which were rejected because there existed established independent unions in these two industries). The final list of industries, along with the number of companies providing information on productivity and injury growth rates of establishments or departments within establishments, the average number of operation and maintenance workers across establishments per year, and the years of survey information, are as follows: textiles (4, 2,000, 1922–5); lumber (13, 2,300, 1923–5); furniture (12, 3,800, 1921–5); paper (15, 7,400, 1921–5); petroleum (1, 3,500, 1921–5); leather (3, 1,700, 1921–5); electrical (1, 7,500, 1921–5); and rubber (1, 1,500, 1921–5).

The data on the number of establishments by industry group in manufacturing is preceded by a lengthy discussion of the various kinds of producers that inhabit each group. This allowed for significant consistency in the assignment and creation of industry categories across the three data sets. However, there was no attempt by the American Engineering Council to make the survey of productivity and injury rate trends representative of the specific industry experiences. For example, one firm (with perhaps a number of plants), producing "electrical machinery, apparatus and supplies" and employing 7,500, represents the entire experience of the "electrical equipment and supplies" industry group for purposes of this analysis.

The variables used in the analysis are defined as follows: *productivity* is the percentage change in the annual output (in physical units—e.g., tons, cubic feet, pounds, etc.) per 100 hours of labor over the period (typically) 1921–5, and is weighted by employment proportions (when necessary) for the creation of industry averages; *injuries* is the percentage change in the annual number of lost-time injuries per million hours of labor over the period (typically) 1921–5, and is weighted by employment proportions (when necessary) for the creation of industry averages; *company union* is the (average) number of company unions in an industry (for 1922 and 1924) divided by the number of establishments in that industry in 1923 multiplied by 100; *horsepower/worker* is the percentage change in horsepower (measured capacity of prime movers and motors) per wage earner by industry from 1923 to 1925 multiplied by 2 (data for 1921 were not available); *company union rank* is the ranking from 1 to 8 (1 being the highest) on the company union index; *cooperative shopfloor outcome rank* is a ranking from 1 to 8 (1 being the best) on the extent of joint improvements in productivity and safety (joint improvements always trump partial improvements in the ranking—for example, lumber ranked fourth with a productivity rate increase of 10.1 percent and an injury rate decline of 12.4 percent, while leather ranked fifth with a productivity rate decrease of 3.2 percent and injury rate decrease of 32.6 percent—and productivity and injury rate performances are weighted equally in determining the ranking).

The data for the analysis in Table 1.2 and thereafter come from the following sources: the injury rate data for 1925 are from a Bureau of Labor Statistics Bulletin (US Department of Labor 1929: 18–20), and for 1926–8, from the *Monthly Labor Review* (US Department of Labor 1930: 56–8). The company union concentration index was created using National Industrial Conference Board data on the number of establishments with company unions across industries in 1924 (National Industrial Conference Board 1925: 8, Table 3), and data on the number of establishments by industry category in 1925 from the *Historical Statistics of the United States: Colonial Times to 1970* (US Bureau of the Census 1975: 669–80). Data on horsepower, value of products, the number of wage earners, and wholesale price indexes by industry are from the *Statistical Abstract of the United States 1932* (US Department of Commerce 1932: 733–85, 299–303). Total factor productivity measures by industry are from Kendrick (1961: 468–75, Table D-IV). Real wage data by industry are from Beney (1936: 60–151).

The problem of an inexact classification of industries in combining the various data sets continues to be a problem in this analysis as it was in earlier analyses. The injury and productivity data are much more problematic in this analysis compared to our analysis of the early 1920s. The injury rate data do not come from the same set of establishments over time; the sample changes each year and the average industry injury rates for the period have rather high variances. The wholesale price indexes by industry are not terribly reliable for this period (our price index for "electrical machinery," for example, is composed of the price of washing machines), lending some concern to our the measure of labor productivity.

The variable definitions are as follows: *productivity* is the percentage change in the real value of products (value of products divided by the wholesale price index) per wage earner from 1925 to 1929; *injuries* is the percentage change in the average number of lost-workday accidents per million employee hours from 1925–6 to

1927–8 (in the four instances in which data were not available for 1925, the comparison was 1926–7 to 1927–8); *company union* is the number of company unions in 1924 divided by the number of establishments in each industry in 1925; *horsepower/worker* is the change in horsepower (measured capacity of prime movers and motors) per wage earner by industry from 1925 to 1929; *company union rank* is the ranking from 1 to 12 (1 being the highest) on the company union index; *cooperative shopfloor outcome rank* is a ranking from 1 to 12 (1 being the best) on the extent of joint improvements in productivity and safety (joint improvements always trump partial improvements in the ranking, and productivity and injury rate performances are weighted equally in determining the ranking); *total factor productivity* is a measure of the change in the real value of output divided by the real value of inputs by industry between 1919 and 1929; *wages* is the change in real hourly earnings of wage earners by industry between 1920 and 1929.

CHAPTER 3

The data for the analysis presented in Table 3.1 are from the following sources: injury rates are from various issues of the *Monthly Labor Review* (US Department of Labor, 1939a: 602–7; 1939c: 874–6; 1940b: 92–4; 1941b: 334–8); union concentration comes from Peterson (1942: 6); strike intensity is the average number of strikers over the period gathered from various issues of *Monthly Labor Review* (US Department of Labor, 1938: 1,190–2; 1939b: 1,114–6; 1940a: 1,090–92; 1941a: 1,097–9), divided by the average number of wage earners for 1937 and 1939, from the *Statistical Abstract of the United States, 1942* (US Department of Commerce 1943: 891–918); productivity growth involves the value of output, the wholesale price index, and the number of wage earners—all from the *Statistical Abstract of the United States, 1942* (1943: 891–918, 430–4, and 891–918, respectively)—and worker-days idle due to strikes, from the *Monthly Labor Review* (US Department of Labor, 1938: 1,190–2; 1940a: 1,090–2).

The data for the injury rate analysis are from the following industries: automobiles, glass, electrical machinery, iron and steel, machinery (excluding agricultural and electrical), rubber, shipbuilding, sugar refining, baking, cement, furniture, leather, meat packing, pottery, boots and shoes, woolens and worsted, canning and preserving, chemicals, confectionery, cotton textiles, flowers and grain, lumber, pulp and paper, and silk and rayon. The data for the productivity analysis are from automobiles, iron and steel, rubber, cement, furniture, leather, boots and shoes, woolens and worsted, chemicals, cotton textiles, lumber, pulp and paper, and silk and rayon.

The inexactness of the system of industrial classification remains a problem both in merging different data sets and in tracking variables over time. There were confusing changes, for example, in industrial categories between 1935 and 1937. This was one of the reasons prompting me to concentrate on the period after 1935. Another was the fact that injury rate statistics improve dramatically beginning in 1936. The categorical measure of union concentration is less than satisfying.

The variable definitions are as follows: *injuries* is the percentage change in the number of lost-workday accidents per million hours worked from 1936–7 to 1939–40 (for the injury rate analysis) and from 1937 to 1939 (for the productivity rate analysis); *union* is a categorical index of the proportion of wage earners under written union agreements in 1941, where 1 indicates the lowest proportion and 4 indicates the highest; *strike intensity* is the average percentage of the labor force involved in strikes during the period 1936 to 1940; and *productivity* is the percentage change in the real value of output divided by the number of worker-days between 1937 and 1939 (worker-days are computed as the number of wage earners multiplied by 260 days minus the number of worker-days idle due to strikes).

CHAPTER 5

The data for the analysis presented in Tables 5.1 and 5.3, and Figures 5.1 and 5.2, are from the following sources: US manufacturing injury rates are from various Bulletins of the Bureau of Labor Statistics (US Department of Labor, 1953: 42; 1954a: 46; 1972: 58) and the *Monthly Labor Review* (US Department of Labor, 1954b: 1,227; 1956: 59; 1957: 64; 1958: 56; 1959: 47); manufacturing injury rates in Japan are from the *Japan Statistical Yearbook* (Bureau of Statistics, 1954: 510; 1957: 508; 1958: 516; 1959: 522; 1960: 520; 1961: 528; 1962: 532; 1963: 538; 1964: 544; 1965: 548; 1966: 634; 1967: 648; 1968: 640; 1969: 637; 1971: 633); manufacturing injury rates in Germany are from *Arbeitsunfallstatistik für die Praxis* (Hauptverband der gewerblichen Berufsgennossenschaften, 1979: 21); industry production indexes used to control for the level of economic activity in the results of Table 5.1 are from *Historical Statistics of the U.S.* (US Bureau of the Census 1975: 667–8); shopfloor power comes from a survey of labor historians and industrial-relations scholars; total factor productivity growth by industry comes from Kendrick and Grossman (1980: 141–59); unionization rates are from Lewis (1963: 274) and Freeman and Medoff (1979: 143–74); average real hourly earnings data by industry come from *Employment and Earnings, U.S. 1909–75*, Bulletin 1,312–10 (US Department of Labor, 1976: 1,212–18) and *Monthly Labor Review* (US Department of Labor, Bureau of Labor Statistics 1972: 114).

The precise definitions of variables used in the analyses are as follows: *Injuries* in the US is the number of disabling work injuries (causing one or more day's absence from work) per million employee hours, and changes in injuries (for Table 3.1 results) are annual average percentage changes over the period; Japanese injury rates are similarly defined, but are gathered only from firms with over 100 workers; German injury rates, measured as the number of injuries per 10,000 workers, are available in ten-year intervals only and were calculated as an employment-weighted average of the injury frequency rates in eight major manufacturing industries because Germany does not report injury rates for manufacturing as an aggregate; *Injury Rate Index* is the ratio of the injury frequency rate in a particular year to the injury frequency rate in the base year (1952 for Japan, 1950 for Germany, and 1948 for the US); P_{it} is an index of productive activity; *% union* is the percent of production workers in the industry covered by collective-bargaining agreements; *shopfloor* *power* is defined and described in the text; *U-shape* is a dichotomous variable, equaling 1 if the industry displayed a U-shaped pattern in its injury rate trajectory over the postwar period and 0 otherwise; *productivity* is the annual average percentage change in total factor productivity over the period; *productivity increase* is a dichotomous variable equaling 1 if the industry witnessed an increase in productivity growth between 1948–60 and 1960–6, and 0 otherwise; *productivity decrease* is a dichotomous variable equaling 1 if the industry witnessed a decrease in productivity growth between 1960–6 and 1966–9, and 0 otherwise; earnings were measured as the real gross average hourly earnings of production workers.

CHAPTER 7

The data for the analysis in Table 7.1 come from two sources: injury and CTD rates are from various Bulletins of the Bureau of Labor Statistics (US Department of Labor, 1988: 26–40, 46–9; 1989: 20–34, 40–3; 1990: 23–36, 43–6; 1991: 23–37, 44–7; 1992: 23–36, 42–5; 1993: 23–6, 46–8). The remaining data come from the Organization of Work in America Business Survey. This is a survey of establishments with 50 or more workers based on the frame maintained by Dun's Marketing Services and conducted by the Center for Survey Research at the University of Massachusetts at Boston.

The precise definitions of variables used in the analysis are as follows: *injuries* is the annual average change in the lost-workday injury rate (injuries per 100 workers) from 1986 to 1991 for the 3-digit manufacturing industry group to which each establishment belongs; *CTD* is the annual average change in the cumulative trauma disorders rate (CTDs per 10,000 workers) from 1986 to 1991 for the 3-digit manufacturing industry group to which each establishment belongs; *QC* is the percentage of the establishment's core workforce involved in quality circles; *teams* is the percentage of the core workforce involved in self-directed work teams; *TQM* is the percentage of the core workforce involved in total quality management; *rotation* is the percentage of the core workforce involved in job rotation; *skill change* is a dichotomous variable equaling 1 if the establishment reported that the skills involved in doing the jobs of the core workforce had changed in the past few years, and 0 otherwise; *plant size* is the number of employees at the establishment; *years experience* is the number of years that the establishment has possessed quality circles; *% union* is the percentage of eligible blue-collar workers covered by collective-bargaining agreements; *% female* is the percentage of the plant workforce that is female; *% blue collar* is the percentage of the plant workforce composed of blue-collar workers.

NOTES

INTRODUCTION

1 The following section is based on an interview on May 23, 1996 with Joe Buckley, shop chairman of Local 696 at the Dayton GM brake plant.

1 FROM EXIT TO VOICE IN SHOPFLOOR GOVERNANCE

1 It should be noted, though, that factories of over 1,000 workers represented less than 20 percent of the manufacturing labor force at this time (Granoveter 1984: 326).

2 The large proportion of inexperienced immigrant labor was a contributing factor cited by the study. Foreign-born workers accounted for roughly half of all less-skilled labor in manufacturing at the turn of the century. Their prevalence in such mass-production industries as steel and meat packing was at least this great (Jacoby 1985: 32).

3 The workforce is derived by dividing the total labor hours by 3,000, the number of hours on average that a full-time worker might be expected to work per year.

4 Recent evidence that the average length of job spells during this period was rather high is not inconsistent with high rates of labor turnover (Carter and Savoca 1990; Sundstrom 1988). Very high turnover was restricted to a subset of jobs in the plant—those with particularly onerous conditions, and populated by younger, less-skilled workers.

5 Turnover is defined here as the number of separations relative to the average labor force of each plant (calculated as the average of the minimum and maximum labor force over the period).

6 For skeptical views of the high costs of training less-skilled workers, see Jacoby (1985) and Raff (1988).

7 The *Wingfoot Clan* (December 25, 1915) is the Goodyear company newspaper, which is housed at the company archives in Akron, Ohio.

8 *Wingfoot Clan*, February 18, 1918.

9 Ibid. March 6, 1918.

10 The data appendix contains the precise definitions and the source for these variables. The ten industries for which data could be constructed were autos, clothing, food products, iron and steel, printing and publishing, textiles, paper, chemicals, leather, and rubber.

11 These and other results from the statistical explorations discussed in this chapter are presented more fully in Fairris (1995a).
12 This follows from the Tiebout model of local public finance (Tiebout 1956: 416–24).
13 The other two periods were 1901–4 and 1934–41.
14 See the data appendix for the precise definitions and sources of variables. The only industries for which data could be collected were food products, iron and steel, textiles, paper, leather, and rubber.
15 See North (1990) for an interesting analysis of the role ideology plays in miti-, gating the free-rider problem associated with collective action towards institutional change.
16 See the data appendix for the precise definitions and sources of variables. The six industries for which date could be collected were food products, iron and steel, textiles, paper, chemicals, and leather.
17 *Wingfoot Clan*, November 1, 1919.
18 Ibid. May 3, 1919; October 6,1920.
19 Statistics for the 1920s are calculated by finding the average turnover rate of firms in the sample and reporting the median of those rates (Berridge 1927: 9–13). Statistics for the 1910s are calculated by finding the average (weighted by number on payroll) across all firms (Brissenden and Frankel 1920: 36–56). The weighted average will be larger than the median if larger firms have higher turnover rates.
20 A contrasting view is that high rates of manufacturing unemployment, due largely to stagnant employment growth in manufacturing, were responsible for the decrease in voluntary quits during the 1920s (Slichter 1929; Jacoby 1983: 261–82). Lazonick (1990) argues that this view fails to acknowledge that the growth in labor demand was quite brisk among many of the dynamic mass-production firms of the decade and that the queuing for jobs, in the latter firms especially, may be a sign of increased attractiveness of employment in those firms relative to the alternatives.
21 *Wingfoot Clan*, January 11, 1921; June 26, 1924; July 30, 1924; March 10, 1926.
22 Gullett and Gray report that in eight of their nine case study firms containing company unions in the 1920s, personnel departments were either initiated or greatly expanded in scope upon the formation of a company union (Gullett and Gray 1976: 95–101).
23 Members of the personnel-management movement were fond of citing statistics on the cost savings to be had for firms that reduced labor turnover. In one firm, a 75 percent reduction in turnover purportedly yielded a 30 percent increase in labor productivity (Fisher 1917). In another, a 150 percent reduction in turnover yielded a 10 percent reduction in manufacturing costs and a 42 percent increase in production (cited in Ross 1958: 912).
24 *Wingfoot Clan*, January 20, 1926.
25 Ibid. March 10, 1926.
26 Certainly other factors were at work in this decline in injury rates. US Steel—the largest firm in the industry—did not even possess company unions during these years. It did, however, possess a form of employee representation over issues of workplace health and safety. Safety committees of managers and workers were an important part of the so-called "safety movement"—a movement that US Steel was largely responsible for launching in the first decade of the twentieth century. The safety movement in most firms included safety committees, inspectors, protective devices on dangerous machinery, and in-house medical services.

Goodyear's efforts were another of the early leaders. The "Safety First" campaign was announced with great fanfare in November of 1913. Efforts had been made over the previous few months to guard exposed gears and cover over belts and chains; the "Safety First" campaign, on the other hand, was a move to enlist workers' insights on the matter of improving health and safety (*Wingfoot Clan*, November 8, 1913). Each department would have four workers, appointed for two month terms, devoted to inspecting department safety hazards for two hours per week. These department-level sub-committees would then make reports to the chairs of nine existing safety committees operating in the plant (*Wingfood Clan*, November 8, and November 15, 1913). The November 14, 1914 issue of the *Wingfoot Clan* announced that the number of accidents had fallen by a third since the inception of the "Safety First" campaign one year earlier. By February of 1915, some 500 workers had served on "Safety First" sub-committees (*Wingfoot Clan*, February 20, 1915).

Another important factor in the declining injury rates of the period was workers' compensation legislation. Most of the northern industrial states passed workers' compensation acts during the 1910s. However, few states made it compulsory for firms to insure, and so only New York, Ohio, and Illinois established compulsory systems prior to 1920. But for those insured firms in the 1920s whose premiums varied with accident experience, workers' compensation provided a strong monetary incentive to increase industrial safety. Company unions complemented any such efforts to reduce premiums by identifying areas for, and recommending the form of, safety improvements.

27 See the data appendix for a description of the data. For a more complete presentation of the empirical results, see Appendix Table 1.1.
28 Interestingly, the introduction of a variable capturing changes in horsepower per worker over the early 1920s into the productivity growth equations (rows 1 and 2) left the results largely unchanged.
29 Nonparametric tests may have an advantage over parametric tests, such as regression techniques, when certain features of the analysis—for example, small sample size—make it likely that standard parametric assumptions are violated. We have utilized a nonparametric test in this case, however, because the simple summation of injury rate and productivity growth performance to arrive at a measure of joint shopfloor benefits would not allow us to distinguish cooperative outcomes from those that are noncooperative (i.e., ones which contain dramatic benefits for one party but losses for the other). Under our ranking system, industries with modest improvements in productivity and safety receive a higher ranking than those with, for example, dramatic productivity growth and slight reductions in safety.
30 See the data appendix for the precise definition of variables and their sources.
31 The American Management Association was founded in 1923 to foster this new approach to personnel administration. The Special Conference Committee was initiated in 1919 by ten of America's leading corporations in order to coordinate policies on employment practices. The members of the committee represented the progressive employers of the period who were committed to combating unions through liberal labor policies rather than the open-shop drive of the American Plan. They included Bethlehem Steel, General Electric, Goodyear, International Harvester, and Standard Oil of New Jersey, among others. These were among the firms possessing a firm commitment to company unions over the decade of the 1920s.
32 See the data appendix for a complete description of the data. For a more detailed presentation of the empirical results, see Appendix Table 1.2.

Alternative measures of labor productivity were considered, but they resulted in a much reduced sample size and generally left our reported findings unchanged. Using Fabricant's (1940) measure of output by industry (as opposed to value of output which is used here) to generate the productivity data yielded a drastically reduced sample size. Replacing the number of wage earners in the denominator of our productivity measure with an index of worker hours (from Beney 1936) reduced the sample size by a quarter and yet yielded similar results to those presented here.

33 It might be claimed that this comparison is not legitimate because the two samples are composed of different industries. However, the results are not substantively altered by restricting the analysis of productivity growth to the six industries that are common to both samples. The coefficient on the company-union index for this analysis is positive and significant in the early 1920s and positive but insignificant and smaller in size in the late 1920s.

34 A second possible source of bias in the results—the omission of a variable capturing average establishment size, which is arguably positively related to company-union concentration and negatively related to workplace hazards— proved to be insignificant. None of the injury rate results were affected by the addition of a control variable for the average number of full-time workers per establishment across industries in 1927. Establishment size was negative and significant (.10 level) in a simple regression of the change in industry injury rates, but became insignificant once the company-union index and other variables were added to the injury rate equation.

35 Once again, the introduction of the mechanization variable into the productivity growth equation left the results largely unchanged.

36 The average rate of change in horsepower per worker over the two periods was 19 percent for our sample of industries. If the rate of change in mechanization had been 19 percent during the period 1921–5, a 1 percent increase in company-union concentration in an industry would have led to a 10 percent decrease in the injury rate. For a similar rate of change in mechanization in the 1925–9 period, however, a 1 percent increase in the company-union coverage would have led to a 3 percent reduction in the injury rate. Again, it might be claimed that this comparison of injury rate trends in the early versus late 1920s is not legitimate because the two samples are composed of different industries, thereby possibly giving rise to different results even without any structural change in the ability of company unions to affect shopfloor safety. However, the results are not substantively changed when the analysis is restricted to the six industries that are common to the two samples. The coefficient on the company-union index in the simple regression of injury rate changes is negative and significant in the early 1920s and negative but insignificant and smaller in absolute value in the late 1920s.

37 Changes in total factor productivity measure real enhancements in the productive efficiency of labor and capital combined. Such measures are not influenced by substitution between inputs in production as are the partial measures of productivity growth—the change in output divided by labor input—used in our earlier analyses. Thus, the change in total factor productivity is a superior measure of productivity growth. Unfortunately, measures of total factor productivity are not available for subperiods of the 1920s, only for the decade as a whole. See the data appendix for the definition and source of this variable.

38 The eight industries are chemicals, leather, paper, petroleum, furniture, electrical machinery, rubber, and food.

39 More nuanced contemporary treatments of company unions can be found in

Brody (1980) and Nelson (1982a). Paul Douglas's 1921 article is a model of clarity in distinguishing between the efficiency and distributional consequences of worker voice. In Douglas's view, company unions could heighten production through "much improved plant morale" and "greater individual effort," while independent unions could ensure that labor received its fair share of the gains (Douglas 1921: 104).

2 THE AMOSKEAG PLAN OF REPRESENTATION

1 I have relied on three very useful studies of the Amoskeag Manufacturing Company: Hareven's *Family Time and Industrial Time* (1982) offers useful details on the changing nature of wages, hours, and conditions, and Hareven and Langenbach's *Amoskeag* (1978) offers unique insights into the nature of work and labor–management relations in the mills through interviews with workers and managers. Creamer and Coulter's analysis of the company's decline in *Labor and the Shut-Down of the Amoskeag Textile Mills* (1971), funded by the Work Projects Administration and completed in 1939, presents an extremely thorough account of the company's competitive problems during the 1920s as well as the impact on workers and community of its feeble efforts to address those problems and its eventual shut-down in the mid-1930s. The information on injuries, as well as further insights into the structural features, workings, and types of matters brought before committees of the plan of representation were culled from company files housed in Baker Library at the Harvard Business School.

2 Amoskeag Manufacturing Company, "Turnover of Labor: 1913, 1915," documents housed at Baker Library, Harvard University.

3 Workers at the Amoskeag were involved in mass rallies and other gestures of support for the striking workers of Lawrence, and became restive themselves— engaging in short-lived work stoppages, for example—during the first few months of 1912 (Creamer and Coulter 1971: 173–4).

4 Amoskeag Manufacturing Company, File C-28.

5 Ibid., "Employee Representation – Suggestion 1923–30," File CD.

6 Ibid., File v.C1.

7 Ibid., File v.C3.

8 Ibid., File v.C29.

9 Ibid., File CR5.

10 Company documents also contain an account of all accidents during this period, whether or not they resulted in a lost workday. A similar pattern of abrupt decline in injuries beginning in 1924 occurs in this series as well.

11 Company employment records generally counted personnel as employed even if they were on temporary layoff, as long as the layoff did not exceed three months in duration (Creamer and Coulter 1971: 322).

12 Other explanations for the reduction in the injury rate cannot be ruled out, of course. However, the company's acquisition of new technology during this period, which might arguably have made work less dangerous, appears to have been largely insignificant (Creamer and Coulter 1971: 16–25). Moreover, there is no indication that the process by which accidents were recorded was altered in any way.

13 Amoskeag Manufacturing Company, File v.C5.

3 THE RISE OF AN EMPOWERED SHOPFLOOR VOICE

1 Another historic source of shopfloor discontent for workers in manufacturing—namely, injury rates—declined during these years, falling by roughly 4 percent between 1930 and 1933. Injury rates tend to be pro-cyclical, however, falling with declines in economic activity and rising as economic activity increases. Therefore the effect of the depression on the underlying structure of workplace safety—i.e., holding constant the influence of the level of economic activity—remains unclear.

2 Over half of a survey of firms in 1932 responded that they based layoffs strictly on workers' "efficiency," while less than a fifth reported that seniority was a factor in such decisions (Jacoby 1985: 218).

3 Although insurance and pension plans did not disappear in large numbers among the progressive employers of the period, the proportion of insurance plans funded solely by employers fell significantly, and pension benefits were substantially curtailed during the early depression years (Jacoby 1985: 220). Cohen (1990: 240) notes, for example, that US Steel pension checks fell by as much as 25 percent during this period.

4 Some of the details of this strike were taken from the Federal Mediation and Conciliation Service files in the National Archives; Record Group 25, File 170–6272.

5 48 Stat. 195 (1933).

6 *Schechter Poultry Corp.* v. *United States*, 295 US 495 (1935).

7 *Matter of Houde Engineering Corporation*, National Labor Relations Board, vol. 1, August 30, 1934: 35.

8 National Labor Relations Act, 1935.

9 Case no. C-41, November 12, 1936.

10 Of the 966 secret-ballot representation elections overseen by the NLRB between October of 1935, and January of 1938, "company unions" appeared on the ballot in 212 of them, and won roughly 50 percent of these (Brooks 1939: 70).

11 In electrical manufacturing, where employers had historically maintained a policy of nondiscrimination towards union membership, another form emerged: local unions were started by rank and file workers with little contact with outside organizers or organizations (Schatz 1983: 63).

12 The collective strike and organizing activity of the late 1910s had been built to a significant degree on the strength of ethnic ties among workers. Events in the Chicago steel mills during the 1919 steel strike offer a useful illustration of the role ethnic identity played during the formative stages of the strike (Cohen 1990: 40–2). Poles, Lithuanians, and Serbs were the backbone of the strike effort in the steel mills in the Chicago area. Workers' ties to separate immigrant communities, and associated institutions such as shops, churches, mutual benefit societies, and fraternal organizations, allowed them to counteract free-rider tendencies among workers of similar ethnicity. Poles who continued to work at the Wisconsin Steel plant in South Chicago during the strike, for example, were threatened by striking Polish workers with expulsion from benefit societies, while neighborhood ethnic shopkeepers who continued to sell goods to scabs were threatened with boycotts from strikers.

13 See Gerstle (1989) for a useful account of the government's efforts to Americanize immigrant workers during this period and its impact on tendencies towards the homogenization of working-class culture and political practice.

14 The formal change of name to the Congress of Industrial Organizations did not take place, however, until 1938.

15 Smaller manufacturers were less committed to this approach, and thus witnessed more confrontational tactics by workers (Schatz 1983: 66).

16 Under measured day work, workers received a high hourly wage but were required to meet stringent company production standards. Disputes over payment systems across industries at this time commonly revolved around their impact on work pace and shopfloor social relations among workers. Steel and packinghouse workers, for instance, struggled against the introduction of piece work and bonus pay systems during this period because they were seen by workers as "slave driving" systems (Cohen 1990: 318).

17 Future contract language, which would require that anyone working on production be a union member and conceding the right of workers to strike over production standards after grievances concerning such matters had been carried to the highest level of the grievance procedure (short of arbitration), would begin to formally address some of workers' shopfloor concerns. However, during the late 1930s workers relied primarily on informal measures for shopfloor influence.

18 See the data appendix for the exact definitions and sources of variables. Appendix Table 1.1 offers a more complete description of the empirical results.

19 Ford, Westinghouse, Goodyear, and the Little Steel firms were brought into the CIO fold during 1941, in the midst of war-time production. During the year-and-a-half from mid-1940 to the end of 1941, organized labor gained about a million and a half new recruits, boosting union membership by over a quarter.

20 The denominator of the injury rate measure—million employee hours worked—appears to capture actual hours worked, and not just an estimate of work hours derived, say, from multiplying the industry labor force by the normal length (in hours) of the work week. The latter measure would fail to account for strike activity, thereby yielding a mismeasured injury rate that fluctuates with strike activity and renders problematic our interpretation of the results. The industry injury rate data reveal amazing consistency for the years 1936 and 1937, even though the difference in strike activity for these two years is quite significant.

21 The need for such an adjustment is clear in looking at the data on the number of workers by industry, which is obviously a measure of the industry labor force without any correction for strike activity. See the data appendix for a fuller discussion of the adjustment procedure.

22 Note that in subtracting man-days idle due to strikes from our measure of total worker days (see the data appendix), we are implicitly assuming that strikes always led to a cessation of operation. To the extent striker replacement workers were a significant phenomenon during this period, our measure of productivity is biased upward because actual worker days are larger than the measure of worker days we employ. Strike activity was generally much greater in 1937 compared to 1939; thus, under such circumstances, our measure of productivity change is biased downward in those industries experiencing significant strike activity over the period, thereby complicating any interpretation of the strike intensity/productivity growth relation. Having accounted for strike intensity, however, the relationship between union concentration and productivity growth should be relatively free of such complications.

4 LABOR–MANAGEMENT DISPUTES IN MEAT PACKING, 1936–41

1 The federal act establishing the Department of Labor granted the Labor Secretary the power to act as a mediator in labor–management disputes. The United States Conciliation Service was thus created within the Labor Department, and a Director of Conciliation was appointed and charged with the responsibility for dispatching commissioners of conciliation to the scene of industrial disputes for purposes of mediation. This agency became the FMCS in 1947, and became independent of the Labor Department at that time.

The files are housed at the National Archives in Washington, DC and are referenced there as FMCS files, Reference Group 280. The files typically contain initial correspondence requesting the services of a federal mediator, and later communications between the commissioners of conciliation and the Director concerning the specifics of the disputes and progress towards their resolution. Sometimes newspaper clippings and labor–management agreements which resolved the disputes are also included in the files.

2 Federal Mediation and Conciliation Service files 1936–41, Record Group 280.
3 The four other sit-down strikes involved disputes over management's refusal to meet with a union grievance committee and workers' refusal to work with workers who expressed hostility towards the union.
4 Federal Mediation and Conciliation Service 1936–41, file 199–2412.
5 Ibid., file 199–1838.
6 Ibid., file 199–4264.
7 Ibid., Don Harris to James Henderson, Secretary, United Packing House Workers of America Local Union, no. 73.
8 Federal Mediation and Conciliation Service 1936–41: file 199–2524.
9 Ibid., file 196–6353.
10 Ibid., T.G. Boughan, Superintendent, to Sidney Hillman, Office of Production Management, July 11, 1941.
11 Federal Mediation and Conciliation Service 1936–41: file 196–6353.
12 Ibid., file 199–7213.

5 INSTITUTIONALIZATION AND DECLINE IN WORKERS' SHOPFLOOR POWER

1 See Stone's (1981) and Tomlins' (1985) discussion of the philosophy of "industrial pluralism" and Brody's (1992) discussion of the system of workplace contractualism.
2 Some scholars of war-time industrial relations maintain that the informal work groups and plant networks of militants formed during the industrial organizing drives had been "atomized" by conscription during the war (Weir 1975). However, both the incidence of and participants in strikes call this analysis into question. Glaberman notes that "Too many of the names that appeared during the organizing days . . . reappeared in the wartime wildcats for [this] thesis to be acceptable" 1980: 34).
3 Brody's earlier analysis of the postwar industrial relations system (1980: 172–214) contains a more subtle treatment of postwar shopfloor governance.
4 Domestic steel producers accounted for 54.1 percent of world production of raw steel in 1946, but by 1970 their share had fallen to 20.1 percent (Hoerr 1988: 94). For the first time in the twentieth century, steel imports exceeded

exports in the US during 1959 when domestic steel mills were closed during the four-month steel strike of that year. For the next twenty years, domestic producers split the growth in domestic steel demand almost equally with importers (Hoerr 1988: 98). In autos, the first domestic Toyota dealership was opened in 1957, while 1960 marked the year of the domestic landing of the Volkswagen Beetle.

5 Material on Goodyear cited in this section is contained in boxes 59 and 61, accession #5583, of the Organization File at the Martin P. Catherwood library at Cornell University.

6 Several of these studies can be found in General Electric Company documents in the Organization File at the Martin P. Catherwood library at Cornell University. The study which is cited above is in Box 54 of the GE files.

7 Henceforth, I refer to these two interchangeably, or merely as automation, since automated technologies are only a more advanced stage of mechanization.

8 There are two broad types of automated technologies that affected production processes in the major manufacturing industries of the postwar period. The first, more common to the automobile and steel industries, is the "transfer technology" type of automated technology, in which a product is produced primarily through automatic transfer between machines that in some cases may also automatically do the processing. The second, more common to the oil and chemical industries, is "continuous process" technologies, in which the level of automation is carried even further to include automatic inspection and feed-back mechanisms with both quality and quantity control.

9 Technological developments alone cannot carry the entire burden of explaining the declining quality of shopfloor conditions during this period, especially in the area of workplace safety. Similar changes in technology were presumably taking place in German and Japanese manufacturing, and yet neither country witnessed the rise in manufacturing injury rates that took place in the US during the 1960s. (See Figure 5.2 and the further discussion below.)

10 See, for example, the USW/US Steel contract of 1947 (p.21) in "Steel" file among the "Collective Bargaining Agreements" files at the Institute of Industrial Relations Library at the University of California, Berkeley.

11 *United Steel Workers* v. *American Manufacturing Co.* (1960) 363 US 564; *United Steelworkers* v. *Warrior and Gulf Navigation Co.* (1960) 363 US 574; and *United Steelworkers* v. *Enterprise Wheel and Car Corp.* (1960) 363 US 593.

12 The bureaucratization of the grievance procedure made grievance handling a complicated task, demanding legal skills not generally possessed by shop stewards. As grievances made their way to higher steps in the procedure, business agents and committeemen—full-time, nonproduction positions financed by the union, or in some cases by the company—took over as representatives for the workers. Decisions at this stage, including whether or not to appeal an earlier ruling, became further removed from the input of production workers.

13 Firm-wide, multi-plant agreements were necessary in order for unions to prevent employer whipsawing, while multi-employer agreements were often forced on unions by employers who were concerned with union whipsawing (Weber 1967; Hendricks and Kahn 1982). The extent of centralization of power could nonetheless differ considerably across union structures. The national unions in autos and steel, for example, played an important role in contract negotiations, overseeing strikes, and administering the grievance procedure, while union decision making was much more decentralized in paper, meat packing, and auto parts. The relative size of the labor union bureaucracy in the United States can be gleaned from Lipset's estimate that in the early 1960s

there were 60,000 full-time, salaried union officials in the US (one for every 300 workers), as compared to 4,000 in Britain (one for every 2,000) or 900 in Sweden (one for every 1,700) (Lipset 1962: 93).

While industry-wide bargaining emerged in informal ways in the 1940s and early 1950s—for example, through pattern bargaining, in which negotiations at one firm set the standard to be followed in proceeding negotiations with other firms in the industry—many industries adopted more formal industry-wide bargaining structures after the mid-1950s. Industry-wide bargaining structures emerged in steel in 1956, for example, when USW president David McDonald—who had assumed the leadership of the steelworkers' union without ever receiving a vote of a single union member in a local, district, or national election after Phillip Murray's death in 1952—assumed the chairmanship of the negotiating committee at the six largest steel producers. In 1959, twelve steel firms collectively entered negotiations with the steelworkers' union using a four-person negotiating committee (Hoerr 1989: 230).

14 *NLRB* v. *Wooster Division of Borg-Warner Corp.* (1958) 356 US 342, 348.
15 *Fireboard Paper Products Company* v. *NLRB* (1964) 379 US 203, 211–12.
16 The employment contract has never been treated in the law as similar to other forms of contract. The modern legal view of employment contracts is influenced by a combination of master–servant notions found in preindustrial law—in which deference, loyalty, and subservience to the employer was stressed—and Taylorist notions of the necessity of managerial control over production for the maintenance of high labor productivity (Atleson 1983). Admittedly, the postwar vision of the employment relationship was significantly different from the nineteenth- and early-twentieth-century legal view, especially in its conception of the employment contract as "a 'charter' or 'code' establishing a system of private law for governing, and an adjudicatory mechanism for resolving, disputes within the work place" (Klare 1978: 249). However, legal opinion has clearly sought to influence the nature of this private "charter," and this influence has contained many of the earlier conceptions of the differential status of labor and management in production.
17 These struggles strained already tense relations between the rank and file and the labor leadership. Contract rejections by workers—an event unheard of before the early 1960s—jumped from 8.7 percent of FMCS "joint-meeting cases" in 1964 to 14.2 percent in 1967 (US Federal Mediation and Conciliation Service 1970: 37). And between 1964 and 1969 significant movements developed to depose union leaders in steel, electrical equipment, and rubber, all of which were successful. This challenge filtered down with even greater vigor to the local level. For example, in the steelworkers' union, new local presidents were elected in 1,100 of the union's 3,800 locals in 1970 (Mkrtchian 1973: 146).
18 In the late 1960s, some international unions—such as the UAW—began authorizing strikes over production standards and health and safety issues during the term of the contract (Lichtenstein 1986: 136).
19 See the data appendix for sources of the injury rate statistics.
20 Manufacturing labor productivity measures reported here were computed from Table 3 of Kendrick (1983), which gives annual averages over business cycle periods.
21 The critical determinant in the new analysis is average hourly wages.
22 See the data appendix for a discussion of the injury rate variables in Figure 5.2 and their sources.

23 While the survey response rate was only 20 percent, the results for those who did respond were highly consistent. There was very little variation across the survey responses on the industry rankings. Moreover, the respondents, as instructed, do not appear to have used the extent of unionization in the industry as a proxy for measured shopfloor power. There was no statistically significant association between the industry average shopfloor power rating and the percent of the industry labor force that is unionized.

24 Two alternative measures of shopfloor power were developed using the survey results: (1) a dichotomous measure, which took on the value 0 or 1 depending on whether the average shopfloor power measure for the industry was less than or equal to 2, or greater than 2, respectively; and (2) a categorical measure derived by rounding off the survey averages for each industry to their nearest whole number. The results using these alternative measures were not substantively different from those reported in the text.

25 See the data appendix for an explanation of the sources and exact definitions of the variables used in the analysis. The more detailed set of findings can be found in Table 5.3.

26 Knowledge of the extent of workers' shopfloor power across industries is not widely held. (Witness the low response rate to our survey.) Knowing more about the extent to which shopfloor power changed across industries during this period will require much more case study research at the plant, firm, and industry level.

27 Changes in the trajectory of industry injury rates might also be attributed to technological developments during these years. Unfortunately, data on horse-power per worker, which would allow us to control for across-industry differences in the rate of automation, were unavailable for much of the period under analysis.

28 The results in rows 1 and 2 are fairly robust to changes in specification. For example, when variables capturing the change in real average hourly earnings and the change in industry output were added to the estimating equation of row 1, neither was statistically significant. Similar results were obtained when these were added to the estimating equation in row 2.

29 This analysis was limited to the seventeen industry categories in Table 5.1. Because separate measures of the level of economic activity were not available for the machinery and the electrical machinery, equipment and supplies indus-tries, we could not test for the existence of a U-shaped pattern in injury rates in those industries. The raw injury rate data for these industries suggests, however, that both would likely have exhibited a U-shaped pattern after accounting for fluctuations in economic activity. Their worker shopfloor power survey score averages were 4 and 3 respectively. Thus, their inclusion would probably have strengthened the findings reported in the results of row 2.

30 Wolff (1985) offers both a useful summary and attempted synthesis of the various studies.

31 Measures of total factor productivity growth capture the joint productivity of capital and labor, and are not sensitive to capital/labor substitution as are labor productivity (i.e., output per worker) measures. The annual average rates of total factor productivity growth over the three periods 1948–60, 1960–6, and 1966–9 are 2.03, 3.89, and 1.26 respectively.

32 In none of the estimated productivity equations to follow was the introduction of an injury rate variable significant in altering the results.

33 The dependent variables in the estimating equations of rows 5 and 6 are dichotomous, implying that a linear probability approach, which does not

restrict predicted values to the 0–1 interval, is inappropriate. However, preliminary estimates revealed a high degree of heteroskedasticity, indicating that neither a logit or probit technique, which does not correct for heteroskedasticity, is appropriate either. We have chosen to present the estimates from a linear probability model which are corrected for heteroskedasticity. The predicted values in both cases were quite reasonably bounded within the 0 to 1 interval, suggesting that the linear probability model is indeed the more appropriate choice.

6 POSTWAR COLLECTIVE-BARGAINING AGREEMENTS

1 The materials for this case study come from collective-bargaining agreements on file at the Institute of Industrial Relations Library at the University of California, Berkeley; the Martin P. Catherwood Library at Cornell University; and the Walter P. Reuther Library at Wayne State University.
2 See, for example, US Steel contracts with the USW in the early 1960s in the "Steel" file, Institute of Industrial Relations Library, University of California, Berkeley.
3 Goodyear, boxes 166–8, Martin P. Catherwood Library, Cornell University; Firestone, "Rubber" file, Institute of Industrial Relations Library, University of California, Berkeley.
4 "Autos" file, Institute of Industrial Relations Library, University of California, Berkeley.
5 Goodrich, "Rubber" file, Institute of Industrial Relations Library, University of California, Berkeley.
6 Meat packing file, Institute of Industrial Relations Library, University of California, Berkeley.
7 Ibid.
8 "Rubber" and "Autos" files respectively, Institute of Industrial Relations Library, University of California, Berkeley.
9 "Meat packing" file, Institute of Industrial Relations Library, University of California, Berkeley.
10 Goodyear, boxes 166–8, Martin P. Catherwood Library, Cornell University.
11 "Autos" file, Institute of Industrial Relations Library, University of California, Berkeley.
12 "UAW Collection," Walter P. Reuther Library, Wayne State University.
13 Ibid.
14 Ibid.
15 "Rubber" file, Institute of Industrial Relations Library, University of California, Berkeley.
16 Ibid.
17 "Steel" file, Institute of Industrial Relations Library, University of California, Berkeley.
18 Ibid.
19 "Autos" file, Institute of Industrial Relations Library, University of California, Berkeley.
20 Ibid.
21 Ibid.
22 Ibid.

7 CONTEMPORARY EXPERIMENTS WITH NEW SYSTEMS OF SHOPFLOOR GOVERNANCE

1 They may also be in violation of labor law's prohibition against any form of worker organization that is dominated by the employer. See the useful discussion in chapter 4 of Gould (1993).

2 This material can be found in Box 54 of the "Company File" at the Martin P. Catherwood Library of Cornell University.

3 Another reason for the demise of these early QWL experiments is waning government support. The Senate held hearings on worker alienation in 1972, and Congress established the National Center for Productivity and Quality of Working Life in 1975 to support QWL experiments. However, the agency was disbanded three years later. It was not until the early 1980s that money was again forthcor. ing by the government (through the Federal Mediation and Conciliation Service) to support participation schemes (Wallace and Driscoll 1981).

4 The surveys of rank-and-file workers occurred in different plants from those of local union officials. The latter surveys took place in five different plants in the auto industry. The former surveys spanned a number of different industries.

5 Similar evidence can be found in more general survey evidence during this period. A national survey in 1985 revealed that 84 percent of workers in firms without a participation program would indeed participate if given the opportunity, and that 90 percent of workers in firms with a participation program reported that the program was a "good idea" (Sirota and Alper Associates 1985: 28).

6 Parker and Slaughter reproduce portions of various local contracts in autos.

7 See, for example, the discussion of NUMMI in Parker and Slaughter (1988), CAMI in Robertson et al. [n.d.], and Subaru-Isuzu Automotive in Graham (1995).

8 Any system of cooperative production will entail some means by which to control the actions of its members so as to ensure the attainment of production targets with minimum waste. The capitalist firm differs from a worker democracy in that the power of stakeholders in the firm is not structurally evenly distributed. The concerns of stockholders and management are disproportionately represented in the structural aspects of the system of control. It is in the actual practice of production that labor expresses its concerns, and, depending on conditions in the labor market and the solidarity of the workforce, practice may diverge from the structural plan. Team production is one element of an emerging structural system of management control.

9 For a fuller discussion of the survey data, see Kochan and Osterman (1994).

10 See the data appendix for the definitions and sources of variables used in the analysis. The year 1986 was chosen as the starting date for calculating injury rates because this was the year that the average establishment in the larger sample had introduced quality circles among its core work group.

11 Appendix Table 7.1 offers the fuller set of findings. Note that the plant size, percent blue collar, and percent female variables are highly statistically significant.

12 Not surprisingly, there is a high degree of collinearity between the various lean-production variables. This gives rise to problems of multicollinearity in the estimated coefficients, which might bias our significance tests. Various attempts to correct for this problem, however, revealed no significant change in the results. Another possible bias in our results stems from selection issues. Worsening injury rates may be a sign of more general problems in the process

of production that could be profitably addressed through TQM techniques. Thus, the introduction of TQM might stem from worsening injuries instead of being their cause. There is an obvious need for further research on this issue.

13 This calculation utilizes the results presented in Appendix Table 7.1.

14 Separate analysis revealed that neither the skill level of the core work group or the percent of part-time workers was significantly associated with the trajectory of injury rates over this period.

15 This measures make use of the findings reported in Appendix Table 7.1.

16 See the data appendix for the definition and sources.

17 Selection issues are, once again, a possible defect in our findings. For example, instead of being a causal factor in the growth of cumulative trauma disorders, the adoption of quality circles could be a response to rising rates of CTDs.

8 A VISIT TO SATURN

1 On the other hand, the ten-hour day, and especially the rotating schedule of day and night shifts to which every worker is assigned, is the source of much worker discontent.

9 THE FUTURE OF US SHOPFLOOR GOVERNANCE

1 The importance of workers' feelings of justice and legitimacy in the productive efficiency of the firm is nicely illustrated in Gouldner's (1954) classic analysis of a wildcat strike and its after effects.

2 Recent surveys revealing the extent of workers' desires for a voice in workplace issues offer further evidence of this claim (e.g., Freeman and Rogers 1993). Workers in lean-production plants express similar desires for a greater say in shopfloor decision making (Robertson et al. n.d.).

3 Interestingly, many Japanese transplants appear to mimic the American version of lean production as well (see Milkman 1991).

4 Senior executives and managers, representing perhaps 2 percent of all employees, are not represented on works councils (Muller-Jentsch 1995: 56).

5 It is estimated that 85 percent of establishments with over 2,000 employees possess "works agreements" that regulate matters not covered by statutory co-determination rights (Muller-Jentsch 1995: 61).

6 Works-council power varies across industries in Germany. In the steel industry, for example, works-council power is quite significant. This is in part due to the stronger co-determination rights at the company level in the coal and steel industries as compared with other industries in Germany (Markovits 1986), but it is also related to the character and extent of union organization among the industry workforces. Works-council power is also significant in autos, where union organization is strong and there exist strong shop-steward committees, compared to, say, electronics, where neither of these exists (Thelen 1991).

7 Centralized bargaining emerged during the 1950s in Germany. Since then there has been very little difference in contract language even across regionally negotiated contracts (Streeck 1984: 13).

8 This section draws heavily on Thelen (1991).

9 A second demand emerged around the issue of wages. Employers had been able to cut wages unilaterally during the downturn in economic activity because many works councils had bid wages up above the levels collectively bargained

in central wage negotiations. Workers reacted to wage cuts by demanding greater decentralization in wage negotiations—for example, that plant-level agreements be reached in a second round of negotiations following central wage bargaining. This would allow contractual protection for the "wage drift" that was before only informally agreed to by the company. The unions gave little formal ground on this front, although the rank-and-file demands resulted in a bit more regional autonomy, shorter contract periods, and a concerted effort by the unions to win sizeable wage increases commensurate with the economic recovery beginning in the late 1960s (Markovits 1986: 206).

10 There is debate concerning the interpretation of works-council rights in the area of health and safety (Gevers 1983), and some indication that works councils are unable (some would say unwilling) to make use of the statutory rights they possess (Wokutch 1990; Gevers 1983). Foremen and supervisors are viewed as the ultimate arbiters of the trade-off between safety and productivity, and works-council meetings are often the site of worker complaints that shopfloor management devalues safety in relation to productivity.

11 Turner cites the foreword to the general principles of group work adopted by the works council at the Wolfsburg VW plant, a portion of which follows: "We need to develop a democratic work culture to show the way for modern democracies; our task is to meet world market risks with our own strengths, those that emerge from a democratic firm culture [as at VW] based on social progress. Good performance and top quality do not come in the long run from pressure or incentives but from interesting work, good teamwork, and appropriate opportunities for input" (Turner 1991: 123, footnote 44).

12 It is sometimes claimed that German works councils are becoming more and more remote from the workers they are meant to represent (Summers 1987), and are increasingly unable to engage in informal, on-the-spot problem solving (Wever 1995). Thus, shopfloor participation groups, such as quality circles and teams, could act as decentralizing mechanisms for shopfloor dispute resolution, thereby fostering those qualities that works councils appear to be increasingly lacking.

13 See Weiler (1990: 282–95) for a description of what I consider to be a minimalist works-council system covering nonshopfloor conditions of employment.

14 Joint labor–management health and safety committees have been mandated by several state-level and province-level governments in the US and Canada. While not strictly comparable to a works-council system with co-determination rights over health and safety, the experience in Canada appears to be quite positive. Workers do not appear to have abused their right to shut down dangerous operations, and employers report benefiting from worker participation and voice in setting policy over workplace health and safety (Adams 1985; Bernard 1995).

BIBLIOGRAPHY

Aaker, D. (1994) "Building a Brand: The Saturn Story," *California Management Review* 36, 2: 114–33.

Abernathy, W. J., Clark, K. B., and Kantrow, A. M. (1983) *Industrial Renaissance: Producing a Competitive Future for America*, New York: Basic Books.

Adams, R. J. (1985) "Should Works Councils Be Used As Industrial Relations Policy?" *Monthly Labor Review* 108: 25–9.

Alexander, M. W. (1916) "Hiring and Firing: Its Economic Waste and How to Avoid It," in *Annals of the American Academy of Political and Social Science* 65 (May): 128–44.

Allen, H. (1949) *The House of Goodyear*, Cleveland: The Coday & Gross Company.

American Engineering Council (1928) *Safety and Production*, New York: Harper and Brothers.

Appelbaum, E. and Batt, R. (1994) *The New American Workplace*, Ithaca: ILR Press.

Aronowitz, S. (1973) *False Promises*, New York: McGraw-Hill Book Company.

Atack, J. and Bateman, F. (1991) "Louis Brandeis, Work and Fatigue at the Start of the Twentieth Century: Prelude to Oregon's Hours Limitation Law," unpublished mimeo, Department of Economics, University of Illinois.

Atleson, J. B. (1983) *Values and Assumptions in American Labor Law*, Amherst: The University of Massachusetts Press.

Beney, A. M. (1936) *Wages, Hours, and Employment in the United States 1914–1936*, New York: National Bureau of Economic Research.

Berggren, C., Bjorkman, T., and Hollander, E. (1991) *Are They Unbeatable? Report from a Field Trip to Study Transplants, the Japanese Owned Auto Plants in North America*, Stockholm: Royal Institute of Technology.

Bernard, E. (1995) "Canada: Joint Committees on Occupational Health and Safety," in J. Rogers and W. Streeck (eds) *Works Councils*, Chicago: The University of Chicago Press.

Bernstein, I. (1960) *The Lean Years: A History of the American Worker 1920–1933*, Boston: Houghton Mifflin Company.

——(1970) *Turbulent Years: A History of the American Worker 1933–1941*, Boston: Houghton Mifflin Company.

Berridge, W. A. (1927) "Factory Labor Turnover—Two New Monthly Indexes," *Monthly Labor Review* 24, 2: 9–13.

Betheil, R. (1978) "The ENA in Perspective: The Transformation of Collective Bargaining in the Basic Steel Industry," *Review of Radical Political Economics* 10, 2: 1–24.

Blauner, R. (1964) *Alienation and Freedom: The Factory Worker and His Industry*, Chicago: The University of Chicago Press.

219

Bluestone, B. and Bluestone, I. (1992) *Negotiating the Future*, New York: Basic Books.

Bowles, S., Gordon, D. M., and Weisskopf, T. E. (1983) *Beyond the Waste Land: A Democratic Alternative to Economic Decline*, New York: Anchor Press/Doubleday.

Brandes, S. D. (1970) *American Welfare Capitalism: 1880–1940*, Chicago: The University of Chicago Press.

Brecher, J. (1977) *Strike!*, San Francisco: Straight Arrow Books.

Brissenden, P. F. and Frankel, E. (1920) "Mobility of Labor in American Industry," *Monthly Labor Review* 10, 6: 1,342–62.

——(1922) *Labor Turnover in Industry*, New York: The Macmillan Co.

Brody, D. (1960) *Steelworkers in America: The Nonunion Era*, New York: Harper Torchbooks.

——(1964) *The Butcher Workmen*, Cambridge: Harvard University Press.

——(1965) *Labor in Crisis: The Steel Strike of 1919*, New York: J. B. Lippincott Company.

——(ed.) (1980) *Workers in Industrial America: Essays on the Twentieth Century Struggle*, New York: Oxford University Press.

——(1992) "Workplace Contractualism: A Historical/Comparative Analysis," in N. Lichtenstein, D. Nelson, and H. J. Harris (eds), *Industrial Democracy in America: The Ambiguous Promise*, Cambridge: Cambridge University Press.

——(1994) "Section 8(a)(2) and the Origins of the Wagner Act," in S. Friedman, R. W. Hurd, R. A. Oswald, and R. L. Seeber (eds) *Restoring the Promise of American Labor Law*, Ithaca: ILR Press.

Brooks, R. R. R. (1939) *Unions of Their Own Choosing*, New Haven: Yale University Press.

——(1940) *As Steel Goes*, New Haven: Yale University Press.

Brown, C. and Reich, M. (1989) "When Does Union-Management Cooperation Work? A Look at NUMMI and GM-Van Nuys," *California Management Review* 31, 4: 26–44.

Buckley, J. (1996) interview with D. Fairris, May 23, Dayton, Ohio.

Bureau of Statistics (various years) *Japan Statistical Yearbook*, Office of the Prime Minister.

Burton, E. R. (1926) *Employee Representation*, Baltimore: The Williams & Wilkins Company.

Cappelli, P. and McKersie, R. (1987) "Management Strategy and the Redesign of Workrules," *Journal of Management Studies* 24, 5: 441–62.

Carter, S. B. and Savoca, E. (1990) "Labor Mobility and Lengthy Jobs in Nineteenth-Century America," *Journal of Economic History* 50, 1: 1–116.

Chamberlain, N. W. (1948) *The Union Challenge to Management Control*, New York: Harper & Brothers Publishers.

Chandler, A. D., Jr. (1977) *The Visible Hand: The Managerial Revolution in American Business*, Cambridge: Harvard University Press.

Chelius, J. R. (1977) *Workplace Safety and Health. The Role of Workers' Compensation*, Washington, DC: The American Enterprise Institute.

Clark, G. (1987) "Why Isn't the Whole World Developed? Lessons from the Cotton Mills," *Journal of Economic History* 47, 1: 141–73.

Cohen, L. (1990) *Making a New Deal*, Cambridge: Cambridge University Press.

Commons, J. R. *et al.* (1935) *History of Labor In the United States, 1896–1932*, vol. 3, New York: Macmillan and Sons Publishing Co.

Cowdrick, E. S. (1931) "Personnel Practice in 1930," *Personnel Series* 11, New York: American Management Association.

Creamer, D. and Coulter, C. W. (1971) *Labor and the Shut-Down of the Amoskeag Textile Mills*, New York: Arno & The New York Times.

Davis, M. (1986) *Prisoners of the American Dream*, London: New Left Books.

Derber, M., Chalmers, W. E., and Edelman, M. T. (1961) "Union Participation in Plant Decision-Making," *Industrial and Labor Relations Review* 15, 1: 83–101.

Dietz, J. W. (1927) "Status of Personnel Men in the Organization," *Annual Convention Series* 58, New York: American Management Association.

Doeringer, P. B. and Piore, M. J. (1971) *Internal Labor Markets and Manpower Analysis*, Lexington: Heath Lexington Books.

Dohse, K., Jurgens, U., and Malsch, T. (1985) "From 'Fordism' to 'Toyotism'? The Social Organization of the Labor Process in the Japanese Automobile Industry," *Politics and Society* 14, 2: 115–46.

Douglas, P. H. (1921) "Shop Committees: Substitutes For, Or Supplement To, Trades Unions?" *Journal of Political Economy* 29, 2: 89–107.

Eastman, C. (1969) *Work Accidents and the Law*, New York: Arno & The New York Times.

Economist (1993) "America the Super-fit," February 13: 67.

Edwards, Richard (1979) *Contested Terrain: The Transformation of the Workplace in the Twentieth Century*, New York: Basic Books.

Elbaum, B. (1984) "The Making and Shaping of Job and Pay Structures in the Iron and Steel Industry," in P. Osterman (ed.) *Internal Labor Markets*, Cambridge: MIT Press.

Fabricant, S. (1940) *The Output of Maufacturing Industries 1899–1937*, New York: National Bureau of Economic Research.

Fairris, D. (1995a) "From Exit to Voice in Shopfloor Governance: The Case of Company Unions," *Business History Review* 69 (Winter): 493–529.

——(1995b) "Control and Inefficiency in Capitalist Production: The Role of Institutions," *Review of Social Economy* 53, 1: 1–29.

——(forthcoming) "Institutional Change in Shopfloor Governance and the Trajectory of Postwar Injury Rates in US Manufacturing, 1948–1970," *Industrial and Labor Relations Review*.

Faunce, W. A. (1958) "Automation in the Automobile Industry: Some Consequences for In-Plant Social Structure," *American Sociological Review* 23, 4: 401–7.

Fine, S. (1963) *The Automobile Under the Blue Eagle*, Ann Arbor: University of Michigan Press.

——(1969) *Sit-Down*, Ann Arbor: University of Michigan Press.

Fisher, B. (1917) "How to Reduce Labor Turnover," in Bureau of Labor Statistics (ed.) *Proceedings of the Employment Managers' Conference*, Bulletin No. 227, Washington, DC: Government Printing Office.

Fraser, S. (1991) *Labor Will Rule: Sidney Hillman and the Rise of American Labor*, New York: The Free Press.

Freeman, R. B. and Lazear, E. P. (1995) "An Economic Analysis of Works Councils," in J. Rogers and W. Streeck (eds) *Works Councils*, Chicago: The University of Chicago Press.

Freeman, R. B. and Medoff, J. L. (1979) "New Estimates of Private Sector Unionism in the United States," *Industrial and Labor Relations Review* 32, 1: 143–74.

——(1984) *What Do Unions Do?*, New York: Basic Books.

Freeman, R.B. and Rogers, J. (1993) "Who Speaks for Us: Employee Representation in a Nonunion Labor Market," in B. Kaufman and M. M. Kleiner (eds) *Employee Representation: Alternatives and Future Directions*, Madison: Industrial Relations Research Association.

French, C. E. (1923) *The Shop Committee in the United States*, Baltimore: Johns Hopkins University Press.

Friedlander, P. (1975) *The Emergence of a UAW Local*, Pittsburgh: University of Pittsburgh Press.

Fucini, J. and Fucini, S. (1990) *Working for the Japanese: Inside Mazda's American Auto Plant*, New York: Free Press, Macmillan.

Galenson, W. (1960) *The CIO Challenge to the AFL*, Cambridge: Harvard University Press.

Georgakas, D. and Surkin, M. (1975) *Detroit: I Do Mind Dying*, New York: St. Martin's Press.

Gerstle, G. (1989) *Working-class Americanism*, Cambridge: Cambridge University Press.

Gevers, J. K. M. (1983) "Worker Participation in Health and Safety in the EEC: The Role of Representative Institutions," *International Labour Review* 122, 4: 26–42.

Gitelman, H. M. (1988) *The Legacy of the Ludlow Massacre*, Philadelphia: University of Pennsylvania Press.

Glaberman, M. (1980) *Wartime Strikes*, Detroit: bewick/ed.

Golden, C. S. and Ruttenberg H. J. (1942) *The Dynamics of Industrial Democracy*, New York: Harper & Brothers Publishers.

Gordon, D. M., Edwards, R. and Reich, M. (1982) *Segmented Work, Divided Workers: The Historical Transformation of Labor in the United States*, New Rochelle: Cambridge University Press.

Gould, W. B. IV (1986) *A Primer on American Labor Law*, Cambridge: MIT Press.

——(1993) *Agenda for Reform*, Cambridge: The MIT Press.

Gouldner, A. W. (1954) *Wildcat Strike*, New York: Harper Torchbooks.

Graham, L. (1995) *On the Line at Suburu, Isuzu*, Ithaca: ILR Press.

Granoveter, M. (1984) "Small is Bountiful: Labor Markets and Establishment Size," *American Sociological Review*, 49 (June): 323–34.

Gullett, C. R. and Gray, E. R. (1976) "The Impact of Employee Representation Plans Upon the Development of Management–Worker Relationships in the United States," *Marquette Business Review* 20 (Fall): 95–101.

Hareven, T. K. (1982) *Family Time and Industrial Time*, Cambridge: Cambridge University Press.

Hareven, T. and Langenbach, R. (1978) *Amoskeag: Life and Work in an American Factory-City*, New York: Pantheon Books.

Harris, H. J. (1982) *The Right to Manage: Industrial Relations Policies of American Business in the 1940s*, Madison: University of Wisconsin Press.

Hauptverband der gewerblichen Berufsgennossenschaften (1979) *Arbeitsunfallstatistik für die Praxis*, Bonn.

Hendricks, W. E. and Kahn L. M. (1982) "The Determinants of Bargaining Structure in US Manufacturing Industries," *Industrial and Labor Relations Review* 35, 2: 181–95.

Herding, R. (1972) *Job Control and Union Structure*, Rotterdam: Rotterdam University Press.

Hirschman, A. (1970) *Exit, Voice and Loyalty*, Cambridge: Harvard University Press.

Hoerr, J. P. (1988) *And the Wolf Finally Came*, Pittsburgh: University of Pittsburgh Press.

Hyman, R. (1972) *Strikes*, Glasgow: Fontana.

Ichniowski, C. (1986) "The Effects of Grievance Activity on Productivity," *Industrial and Labor Relations Review*, 40, 1: 75–89.

Ichniowski, C., Shaw, K., and Prennushi, G. (1994) "The Effects of Human Resource Management Practices on Productivity," mimeo, Graduate School of Business, Columbia University.

Jacoby, S. M. (1983) "Industrial Labor Mobility in Historical Perspective," *Industrial Relations* 22, 2: 261–82.

——(1985) *Employing Bureaucracy: Managers, Unions, and the Transformation of Work in American Industry, 1900–1945*, New York: Columbia University Press.

Jefferys, S. (1986) *Management and Managed: Fifty Years of Crisis at Chrysler*, Cambridge: Cambridge University Press.

——(1989) "Matters of Mutual Interest: The Unionization Process at Dodge Main, 1933–39," in N. Lichtenstein and S. Meyer (eds), *On the Line*, Urbana: University of Illinois Press.

Jerome, H. (1934) *Mechanization in Industry*, New York: National Bureau of Economic Research.

Kansas City Star (1938a) "An Armour Sit-Down," September 9: 1.

——(1938b) September 10: 1.

——(1938c) September 13: 1.

Katz, H. C. (1985) *Shifting Gears: Changing Labor Relations in the US Automobile Industry*, Cambridge: Cambridge University Press.

Katz, H. C., Kochan, T. A., and Gobeille, K. R. (1983) "Industrial Relations Performance, Economic Performance and QWL Programs: An Interplant Analysis," *Industrial and Labor Relations Review* 37, 1: 3–17.

Katz, H. C. and Sabel, C. F. (1985) "Industrial Relations & Industrial Adjustment in the Car Industry," *Industrial Relations* 24, 3: 295–315.

Kendrick, J. W. (1961) *Productivity Trends in the United States*, New York: National Bureau of Economic Research.

——(1983) *Interindustry Differences in Productivity Growth*, Washington, DC: American Enterprise Institute.

Kendrick, J. W. and Grossman, E. S. (1980) *Productivity in the United States: Trends and Cycles*, Baltimore: The Johns Hopkins University Press.

Klare, K. E. (1978) "Judicial Deradicalization of the Wagner Act and the Origins of Modern Legal Consciousness, 1937–1941," *Minnesota Law Review* 62: 265–339.

——(1981) "Labor Law as Ideology: Toward a New Historiography of Collective Bargaining Law," *Industrial Relations Law Journal* 4, 3: 450–82.

Klingel, S. and Martin, A. (1988) *A Fighting Chance: New Strategies to Save Jobs and Reduce Costs*, Ithaca: ILR Press.

Kochan, T. A., Katz, H. C., and Mower, N. R. (1984) *Worker Participation and American Unions: Threat or Opportunity?*, Kalamazoo: W. E. Upjohn Institute for Employment Research.

Kochan, T. A., Katz, H. C., and McKersie, R. B. (1986) *The Transformation of American Industrial Relations*, New York: Basic Books, Inc.

Kochan, T. A. and Osterman, P. (1994) *The Mutual Gains Enterprise*, Boston: Harvard Business School Press.

Kuhn, J. (1961) *Bargaining in the Grievance Settlement*, New York: Columbia University Press.

Lawler, E. E., III, Ledford, G., and Mohrman, S. A. (1989) *Employee Involvement in America*, Houston: American Productivity and Quality Center.

Lawler, E. E., III, Mohrman, S. A., and Ledford, G. (1992) *Employee Involvement and TQM*, San Francisco: Jossey-Bass.

Lazonick, W. H. (1983) "Technological Change and the Control of Work: The Development of Capital–Labour Relations in US Manufacturing Industry," in H. F. Gospel and C. R. Littler (eds) *Managerial Strategies and Industrial Relations*, London: Heinemann Educational Books.

——(1990) *Competitive Advantage on the Shop Floor*, Cambridge: Harvard University Press.

Leiserson, W. M. (1928) "The Accomplishments and Significance of Employee Representation," *Personnel* 4, 4: 119–36.

Lens, S. (1959) *The Crisis of American Labor*, New York: Sagamore Press.

Lescohier, D. D. (1935) "Working Conditions," in J. R. Commons, *et al.* (eds), *History of Labor in the United States 1896–1932*, New York: Macmillan and Sons Publishing Co.

Levin, D. I. (1995) *Reinventing the Workplace*, Washington DC: The Brookings Institute.

Levine, D. and Tyson, L. (1990) "Participation, Productivity, and the Firm's Performance," in A. S. Blinder (ed.) *Paying For Productivity*, Washington, DC: The Brookings Institute.

Lewis, H. G. (1963) *Unionism and Relative Wages in the United States*, Chicago: University of Chicago Press.

Lichtenstein, N. (1982) *Labor's War at Home: The CIO in World War II*, Cambridge: Cambridge University Press.

——(1985) "UAW Bargaining Strategy and Shop-Floor Conflict: 1946–1970," *Industrial Relations* 24, 3: 360–81.

——(1986) "Reutherism on the Shop Floor: Union Strategy and Shop-Floor Conflict in the USA 1946–70," in S. Tolliday and J. Zeitlin (eds) *The Automobile Industry and its Workers*, Cambridge, England: Polity Press.

——(1989) "The Man in the Middle: A Social History of Automobile Industry Foremen," in N. Lichtenstein and S. Meyers (eds), *On the Line*, Urbana: University of Illinois Press.

——(1995) *The Most Dangerous Man In Detroit*, New York: Basic Books.

Lincoln, J. R. and Kalleberg, A. L. (1990) *Culture, Control, and Commitment*, Cambridge: Cambridge University Press.

Lipset, S. M. (1962) "Trade Unions and Social Structure II," *Industrial Relations* 1, 2: 89–110.

Litchfield, P. W. (1919) *The Industrial Republic*, Akron: Goodyear Tire & Rubber Company.

Livernash, E. R. (1962) "The General Problem of Work Rules," *Proceedings of the 14th Annual Meetings, Industrial Relations Research Association, Madison* 28: 389–98.

——(1967) "Special and Local Negotiations," in J. T. Dunlop and N. W. Chamberlain (eds) *Frontiers of Collective Bargaining*, New York: Harper & Row Publishers.

Lynd, A. and Lynd, S. (1973) *Rank and File*, Boston: Beacon Press.

Mann, E. (1987) *Taking on General Motors*, Los Angeles: Institute of Industrial Relations, University of California, Los Angeles.

Mann, F. C. and Hoffman, R. L. (1960) *Automation and the Worker*, New York: Holt, Rinehart and Winston.

Marglin, S. (1974) "What Do Bosses Do?" *Review of Radical Political Economics* 6: 60–112.

Markovits, A. S. (1986) *The Politics of the West German Trade Unions*, Cambridge: Cambridge University Press.

Marsh, R. M. (1992) "The Difference Between Participation and Power in Japanese Factories," *Industrial and Labor Relations Review* 45, 2: 250–7.

Mathewson, S. B. (1931) *Restriction of Output Among Unorganized Workers*, New York: The Viking Press.

McKersie, R.B. and Klein, J. (1985) "Productivity: The Industrial Relations Connection," in W. J. Baumol and McLennan, K. (eds) *Productivity Growth and US Competitiveness*, Oxford: Oxford University Press.

Meyer, S. (1992) *Stalin over Wisconsin*, New Brunswick: Rutgers University Press.

Milkman, R. (1991) *Japan's California Factories: Labor Relations and Economic Globalization*, Los Angeles: Institute of Industrial Relations, University of California.

Millis, H. A. and Montgomery, R. E. (1945) *Organized Labor*, New York: McGraw-Hill.

Mkrtchian, A. (1973) *US Labour Unions Today: Basic Problems and Trends*, Moscow: Progress Publishers.

Montgomery, D. (1979) *Workers' Control In America*, Cambridge: Cambridge University Press.

——(1987) *The Fall of the House of Labor*, Cambridge: Cambridge University Press.

Muller-Jentsch, W. (1995) "Germany: From Collective Voice to Co-management," in J. Rogers and W. Streeck (eds) *Works Councils*, Chicago: The University of Chicago Press.

Myers, J. (1924) *Representative Government in Industry*, New York: George H. Doran Company.

Naples, M. I. (1981) "Industrial Conflict and Its Implications for Productivity Growth," *American Economic Review* 71, 2: 36–41.

——(1988) "Industrial Conflict, the Quality of Worklife, and the Productivity Slowdown in US Manufacturing," *Eastern Economic Journal* 14, 2: 157–66.

Naples, M. I. and Gordon D. M. (1981) "The Industrial Accident Rate: Creating a Consistent Time Series," unpublished paper, New York: Institute for Labor Education and Research.

National Industrial Conference Board (1933) *Collective Bargaining Through Employee Representation*, New York: National Industrial Conference Board.

——(1934) *Effect of the Depression on Industrial Relations Programs*, New York: National Industrial Conference Board.

——(1925) *The Growth of Works Councils in the United States*, New York: National Industrial Conference Board.

Nauss, D. W. (1994) "Pumping the Brakes: Losing Money, GM is Cautiously Retooling its Strategy for Saturn," *Los Angeles Times* 6 November: D6.

Nelson, D. (1975) *Managers and Workers*, Madison: The University of Wisconsin Press.

——(1982a) "The Company Union Movement, 1900–1937: A Reexamination," *Business History Review* 61, 3: 335–57.

——(1982b) "Origins of the Sit-down Era: Worker Militancy and Innovation in the Rubber Industry, 1934–38," *Labor History* 23, 2: 198–225.

——(1988) *American Rubber Workers & Organized Labor*, Princeton: Princeton University Press.

Niesz, H. E. and Knapp, A. (1927) "The Scope of Activities of a Personnel Department," *Annual Convention Series* 60, New York: American Management Association.

Norsworthy, J. R. and Zabala, C. (1985) "Worker Attitudes, Worker Behavior, and Productivity in the US Automobile Industry," *Industrial and Labor Relations Review* 38, 4: 544–57.

North, D. C. (1990) *Institutions, Institutional Change and Economic Performance*, Cambridge: Cambridge University Press.

North Dakota Union Farmer (1939) September 18: 1.

Olson, M. (1971) *The Logic of Collective Action*, Cambridge: Harvard University Press.

Osterman, P. (1994) "How Common is Workplace Transformation and How Can We Explain Who Adopts It? Results from a National Survey, *Industrial and Labor Relations Review* January: 175–88.

Owen, L. J. (1995) "Worker Turnover in the 1920s: What Labor-Supply Arguments Don't Tell Us," *Journal of Economic History* 55, 4: 822–41.

Parker, M. (1985) *Inside the Circle: A Union Guide to QWL*, Detroit: Labor Notes/South End Press.

Parker, M. and Slaughter, J. (1988) *Choosing Sides: Unions and the Team Concept*, Boston: South End Press.

Peterson, F. (1935) "Unemployment and Relief—Federal and State," in J. R. Commons *et al.* (eds), *History of Labor in the United States 1896–1932*, New York: Macmillan and Sons Publishing Co.

——(1937) *Strikes in the United States, 1880–1936*," US Bureau of Labor Statistics Bulletin No. 651, Washington, DC: Government Printing Office.

——(1942) "Extent of Collective Bargaining at Beginning of 1942," *Monthly Labor Review*, 54 (May), Washington, DC: Government Printing Office.

Piore, M. J. and Sabel, C. F. (1984) *The Second Industrial Divide: Possibilities for Prosperity*, New York: Basic Books, Inc.

Raff, D. M. G. (1988) "Wage Determination Theory and the Five-Dollar Day at Ford," *Journal of Economic History* 48, 2: 387–400.

Robertson, D., Rinehart, J., and Huxley, C. (n.d.) "Team Concept: A Case Study of Japanese Production Management in a Unionized Canadian Auto Plant," CAW Research Group on CAMI, Port Elgin: Ontario, Canada.

Rogers, J. and Streeck, W. (1994) "Workplace Representation Overseas: The Works Council Story," In R. B. Freeman (ed.) *Working Under Different Rules*, New York: Russell Sage Foundation.

——(1995) "The Study of Works Councils: Concepts and Problems," in J. Rogers and W. Streeck (eds) *Works Councils*, Chicago: The University of Chicago Press.

Ross, A, M. (1958) "Do We Have a New Industrial Feudalism?" *American Economic Review* 48, 5: 902–20.

Rubinstein, S., Bennett, M., and Kochan, T. (1993) "The Saturn Partnership: Co-Management and the Reinvention of the Local Union," in B. L. Kaufman and M. Kleiner (eds) *Employee Representation: Alternatives and Future Directions*, Madison: Industrial Relations Research Association.

Sayles, L. R. (1958) *Behavior of Industrial Work Groups*, New York: John Wiley & Sons, Inc.

Sayles, L. R. and Strauss, G. (1967) *The Local Union*, New York: Harcourt, Brace and World.

Schacht, J. N. (1975) "Toward Industrial Unionism: Bell Telephone Workers and Company Unions, 1919–1937," *Labor History* 10, 1: 5–36.

Schatz, R. W. (1983) *The Electrical Workers*, Urbana: University of Illinois Press.

Seidman, J., London, J., Karsh, B., and Tagliacozzo, D. (1958) *The Worker Views His Union*, Chicago: University of Chicago Press.

Sherman, J. (1994) *In the Rings of Saturn*, New York: Oxford University Press.

Sirota and Alper Associates (1985) *The 1985 National Survey of Employee Attitudes*, New York: Sirota and Alper Associates.

Slichter, S. (1919) *The Turnover of Factory Labor*, New York: D. Appleton.

——(1929) "The Current Labor Policies of American Industries," *Quarterly Journal of Economics* 43 (May): 393–435.

——(1941) *Union Policies and Industrial Management*, Washington: Brookings Institution.

Slichter, S. H., Healy, J. J., and Livernash, E. R. (1960) *The Impact of Collective Bargaining on Management*, Washington, DC: The Brookings Institution.

Smith, R. S. (1973) "Intertemporal Changes in Work Injury Rates," *Proceedings of the Twenty-Fifth Annual Meeting*, Madison: Industrial Relations Research Association.

——(1992) "Have OSHA and Workers' Compensation Made the Workplace Safer?," in D. Lewin, O. S. Mitchell, and P. D. Sherer (eds) *Research Frontiers in Industrial Relations and Human Resources*, Madison: Industrial Relations Research Association.

Solomon, C. M. (1991) "Behind the Wheel at Saturn," *Personnel Journal* 70, 6: 72–4.

Spates, T. G. (1937) "An Analysis of Industrial Relations Trends," *Personnel Series* No. 5, New York: American Management Association.

Stone, K. (1981) "The Post-War Paradigm in American Labor Law," *Yale Law Journal* 90, 7: 1,509–80.

Strauss, G. (1962) "The Shifting Power Balance in the Plant," *Industrial Relations* 3: 81.

Streeck, W. (1984) *Industrial Relations in West Germany*, New York: St. Martin's Press.

Summers, C. W. (1987) "An American Perspective on the German Model of Worker Participation," *Comparative Labor Law Journal* 33: 333–55.

Sundstrom, W. A. (1988) "Internal Labor Markets before World War I: On-the-Job Training and Employee Promotion," *Explorations in Economic History* 25 (October): 424–45.

Taylor, F. W. (1919) *Shop Management*, New York: Harper & Brothers Publishing.

Thelen, K. A. (1991) *Union of Parts*, Ithaca: Cornell University Press.

Thimm, A. L. (1980) *The False Promise of Codetermination*, Lexington: Lexington Books.

Tiebout, C. M. (1956) "A Pure Theory of Local Expenditures," *Journal of Political Economy* 64 (October): 416–24.

Tomlins, C. L. (1985) *The State and the Unions: Labor Relations, Law, and the Organized Labor Movement in America, 1880–1960*, Cambridge: Cambridge University Press.

Treece, J. B. (1989) "Shaking Up Detroit," *Business Week*, August 14.

Turner, L. (1991) *Democracy at Work*, Ithaca: Cornell University Press.

US Bureau of the Census (1975) *Historical Statistics of the United States, Colonial Times to 1970*, Washington, DC: Government Printing Office.

US Department of Commerce (1933) *Statistical Abstract of the US, 1932*, Washington, DC: Government Printing Office.

——(1935) *Statistical Abstract of the US, 1934*, Washington, DC: Government Printing Office.

——(1943) *Statistical Abstract of the US, 1942*, Washington, DC: Government Printing Office.

US Department of Labor (1976) *Employment and Earnings, US 1909–75*, Washington, DC: Government Printing Office.

US Department of Labor, Bureau of Labor Statistics (1917) "Strikes and Lockouts in the United States, 1916," *Monthly Labor Review* 4, 4: 600–8.

——(1918) *The Safety Movement in the Iron and Steel Industry, 1907–1917*, Bulletin No. 234, Washington, DC: Government Printing Office.

——(1919) *Welfare Work for Employees in Industrial Establishments in the United States*, Bulletin No. 250, Washington, DC: Government Printing Office.

——(1928) *Health and Recreation Activities in Industrial Establishments, 1926,* Bulletin No. 458, Washington, DC: Government Printing Office.

——(1929) *Statistics of Industrial Accidents in the United States to the end of 1927,* Bulletin No. 490. Washington, DC: Government Printing Office.

——(1930) "Industrial Accidents in the US," *Monthly Labor Review* 30, 1 (January): 56–8.

——(1934) "Industrial Accidents and Safety," *Monthly Labor Review* 38 (May): 1,089–93.

——(1935a) "Report on Conditions in the Automobile Industry," *Monthly Labor Review* 40 (March): 646–7.

——(1935b) *Revised Index of Factory Employment and Pay Rolls, 1919 to 1933,* Bulletin No. 610, Washington, DC: Government Printing Office.

——(1938a) "Strikes During 1937," *Monthly Labor Review* 46 (May): 1,190–2.

——(1938b) *Characteristics of Company Unions 1935,* Bulletin No. 634, Washington, DC: Government Printing Office.

——(1939a) "Industrial Accidents During 1936 and 1937," *Monthly Labor Review* 48 (March): 602–7.

——(1939b) "Strikes During 1938," *Monthly Labor Review* 48 (May): 1,114–16.

——(1939c) "Industrial Accidents During 1936 and 1937," *Monthly Labor Review* 49 (October): 874–6.

——(1940a) "Strikes During 1939," *Monthly Labor Review* 50 (May): 1,090–2.

——(1940b) "Industrial Accidents During 1939," *Monthly Labor Review* 51 (July): 92–4.

——(1941a) "Strikes During 1940," *Monthly Labor Review* 52 (May): 1,097–9.

——(1941b) "Industrial Accidents During 1940," *Monthly Labor Review* 53 (August): 334–8.

——(1950) *Handbook of Labor Statistics,* Bulletin No. 1,016, Washington, DC: Government Printing Office.

——(1953) *Work Injuries in the United States During 1951,* Bulletin No. 1,137, Washington, DC: Government Printing Office.

——(1954) *Work Injuries in the United States During 1952,* Bulletin No. 1,164, Washington, DC: Government Printing Office.

——(1972) *Handbook of Labor Statistics,* Bulletin No. 1735, Washington, DC: Government Printing Office.

——(1988) *Occupational Injuries and Illnesses in the US by Industry, 1986,* Bulletin No. 2,308, Washington, DC: Government Printing Office.

——(1989) *Occupational Injuries and Illnesses in the US by Industry, 1987,* Bulletin No. 2,328, Washington, DC: Government Printing Office.

——(1990) *Occupational Injuries and Illnesses in the US by Industry, 1988.* Bulletin No. 2,366, Washington, DC: Government Printing Office.

——(1991) *Occupational Injuries and Illnesses in the United States by Industry, 1989,* Bulletin 2,379, Washington, DC: Government Printing Office.

——(1992) *Occupational Injuries and Illnesses in the United States by Industry, 1991,* Bulletin 2,397, Washington, DC: Government Printing Office.

——(1993) *Occupational Injuries and Illnesses in the United States by Industry, 1991,* Bulletin No. 2,424, Washington DC: Government Printing Office.

US Department of Labor, Commission on the Future of Worker Management Relations (1994) *Fact Finding Report of the Commission on the Future of Worker Management Relations,* Washington DC: Government Printing Office.

US Federal Mediation and Conciliation Service, (1970) *Annual Report,* Washington, DC: US Government Printing Office.

Walker, C.R. (1957) *Toward the Automatic Factory*, New Haven: Yale University Press.

Wallace, P. A. and Driscoll, J. W. (1981) "Social Issues in Collective Bargaining," in J. Stieber, R. B. McKersie, and D. Q. Mills (eds) *US Industrial Relations 1950–1980: A Critical Assessment*, Madison: Industrial Relations Research Association.

Wall Street Journal (1989) "Workplace Injuries Proliferate as Concerns Push People to Produce," June 16: 8, 41.

——(1994) "9 to Nowhere," December 1: A1, A10.

——(1996a) "How UAW Strike Became Test Ground For GM's Resolve," March 18: 1.

——(1996b) "A Rich Benefits Plan Gives GM Competition Cost Edge," March 21: 1.

Weakly, F. E. (1923) *Applied Personnel Procedures*, New York: McGraw-Hill Book Co.

Weber, A. (1967) "Stability and Change in the Structure of Collective Bargaining," in L. Ulman (ed.) *Challenges to Collective Bargaining*, Englewood Cliffs, New Jersey: Prentice Hall.

Weiler, P. C. (1990) *Governing the Workplace*, Cambridge: Harvard University Press.

Weir, S. (1975) "American Labor on the Defensive: a 1940's Odyssey," *Radical America* 9, 4–5: 163–85.

Weisskopf, T. E., Bowles, S., and Gordon, D. M. (1983) "Hearts and Minds: A Social Model of US Productivity Growth," *Brookings Papers on Economic Activity*, 2: 381–441.

Wever, K. S. (1995) *Negotiating Competitiveness*, Boston: Harvard Business School Press.

Widick, B. J. (1964) "Prototype for More Conflict," *The Nation*, November 16: 349.

Wokutch, R. E. (1990) *Cooperation and Conflict in Occupational Safety and Health*, New York: Praeger.

——(1992) *Worker Protection, Japanese Style*, Ithaca: ILR Press.

Wolff, E. N. (1985) "The Magnitude and Causes of the Recent Productivity Slowdown in the United States: A Survey of Recent Studies," in W.J. Baumol and K. McLennan (eds) *Productivity Growth & US Competitiveness*, New York: Oxford University Press.

Wolman, L. (1924) *The Growth of American Trade Unions: 1880–1923*, New York: National Bureau of Economic Research.

Womack, J. P., Jones, D. T., and Roos, D. (1990) *The Machine That Changed the World*, New York: Rawson Associates.

Work in America (1972) "Report of a Special Task Force to the Secretary of Health, Education, and Welfare," Cambridge: MIT Press.

Zieger, R. H. (1986) *American Workers, American Unions, 1920–1985*, Baltimore: Johns Hopkins Press.

INDEX

230

For Product Safety Concerns and Information please contact our EU
representative GPSR@taylorandfrancis.com
Taylor & Francis Verlag GmbH, Kaufingerstraße 24, 80331 München, Germany

www.ingramcontent.com/pod-product-compliance
Lightning Source LLC
Chambersburg PA
CBHW070400270326
41926CB00014B/2637